Peate's Body Systems

Peate's Body Systems

The Renal System

Ian Peate, OBE FRCN EN(G) RGN DipN(Lond) RNT BEd(Hons) MA(Lond) LLM

Editor in Chief, British Journal of Nursing;
Consultant Editor, Journal of Paramedic Practice;
Consultant Editor, International Journal for Advancing Practice;
Visiting Professor, Northumbria University;
Visiting Professor, Buckinghamshire New University;
Professorial Fellow, University of Roehampton;
Visiting Senior Clinical Fellow, University of Hertfordshire

Registered Office(s)
John Wiley & Sons, Inc., 111 River Street, Hoboken, NJ 07030, USA
John Wiley & Sons Ltd, The Atrium, Southern Gate, Chichester, West Sussex, PO19 8SQ, UK

For details of our global editorial offices, customer services, and more information about Wiley products visit us at www.wiley.com.

Wiley also publishes its books in a variety of electronic formats and by print-on-demand. Some content that appears in standard print versions of this book may not be available in other formats.

Library of Congress Cataloging-in-Publication Data

Names: Peate, Ian, author.
Title: The renal system / Ian Peate, OBE FRCN EN(G) RGN
    DipN(Lond) RNT Bed BEd (Hons) MA(Lond) LLM, Editor in Chief, British
    Journal of Nursing, Consultant Editor, Journal of Paramedic Practice,
    Consultant Editor, International Journal for Advancing Practice,
    Visiting Professor, Northumbria University, Visiting Professor, St
    Georges University of London and Kingston University London,
    Professorial Fellow, Roehampton University, Visiting Senior Clinical
    Fellow, University of Hertfordshire.
Description: Hoboken, NJ, USA : Wiley Blackwell, 2025. | Series: Body
    systems; book 4 | Includes bibliographical references and index.
Identifiers: LCCN 2024041149 (print) | LCCN 2024041150 (ebook) | ISBN 9781394252442
    (paperback) | ISBN 9781394252459 (epub)
Subjects: LCSH: Urinary organs–Physiology. | Kidneys–Physiology. | Kidneys–Diseases.
Classification: LCC QP249 .P43 2025 (print) | LCC QP249 (ebook) | DDC 612.4/6–dc23/eng/20241228
LC record available at https://lccn.loc.gov/2024041149
LC ebook record available at https://lccn.loc.gov/2024041150

Cover Images: © marinashevchenko/Adobe Stock, © phototechno/Getty Images, © 4luck/Adobe Stock, © Oleksandr Pokusai/Adobe Stock
Cover Design: Wiley

Set in 9.5pt STIXTwo by Lumina Datamatics

Printed and bound by CPI Group (UK) Ltd, Croydon, CR0 4YY
C9781394252442_070125

The manufacturer's authorized representative according to the EU
General Product Safety Regulation is Wiley-VCH GmbH, Boschstr.
12, 69469 Weinheim, Germany, e-mail: Product_Safety@wiley.com.

Contents

7 Bladder Cancer 112

8 Urinary Tract Infection 128

Preface

Welcome to *Peate's Body Systems*; there are 12 books in the series. This is a comprehensive collection of textbooks designed to support and enrich the knowledge of health and care workers across various fields. This series is intended to be a valuable resource for those who are dedicated to understanding the intricacies of human biology, physiology and the various systems that sustain life.

Peate's Body Systems series is rooted in the belief that a deep and thorough understanding of the human body is essential for providing the highest standard of care. Each book in this series is thoroughly crafted to offer clear, accurate and up-to-date information on different body systems. The aim is to bridge the gap between complex scientific concepts and practical, everyday applications in healthcare settings.

PURPOSE AND SCOPE

The purpose of this series is to provide health and care workers with:

- Foundational knowledge, with explanations of the anatomical structures and physiological functions of the body systems

- Insights into how these systems interact with each other and how they are impacted by various diseases and conditions, highlighting clinical relevance and encouraging practical application

STRUCTURE OF THE SERIES

Each book in *Peate's Body Systems* focuses on a specific body system:

The Cardiovascular System	The Female Reproductive System
The Respiratory System	The Male Reproductive System
The Digestive System	The Musculoskeletal System
The Renal System	The Skin
The Nervous System	The Ear, Nose and Throat
The Endocrine System	The Eyes

Every chapter is designed to be comprehensive yet accessible, making complex information easier to digest and apply. Figures, tables, boxes, illustrations and flowcharts have been extensively used to support visual learning and reinforce key concepts.

This series is tailored for:

- Healthcare students: those in nursing and allied health programmes

- Practicing professionals: nurses, therapists and other care workers seeking to deepen their understanding and stay current with the latest developments in health and care

- Educators and trainers: educators who require reliable and comprehensive teaching materials to advise and instruct the next generation of healthcare providers

Commitment to Excellence. The series is committed to providing quality educational resources that not only inform but also inspire and empower health and care workers. By equipping you with a robust understanding of the systems of life, you will be better prepared to make informed decisions, deliver compassionate care and ultimately improve patient outcomes.

Thank you for choosing *Peate's Body Systems* as your trusted resource. I hope these textbooks serve as a valuable tool in your ongoing journey of learning and professional development.

IAN PEATE
London

Acknowledgements

I would like to acknowledge the help and support of my partner Jussi Lahtinen. Acknowledgements also go to staff at the RCN Library in London. My thanks go to Tom Marriott, Christabel Daniel Raj, Bhavya Boopathi and all those at Wiley.

Anatomy and Physiology: The Renal System

CHAPTER 1

The renal system is also known as the urinary system; it plays a crucial role in maintaining homeostasis within the body. Comprising the kidneys, ureters, bladder and urethra, this system is responsible for filtering and eliminating waste products from the blood while regulating fluid and electrolyte balance.

The body is an intricately designed collection of systems and organs, relying on the precise coordination of various functions to maintain equilibrium. At the epicentre of this physiological coordination is the renal system, which is a complex network responsible for filtering blood, regulating fluid balance and arranging the expulsion of waste. This chapter explores the renal system, the anatomy, functions and mechanisms that define its role in sustaining the delicate balance of internal homeostasis.

The renal system is located in the lower back, where the two bean-shaped organs house nephrons, the microscopic components of urine formation. The intricate structures and functions of the kidneys are responsible for selective filtration, reabsorption and secretion taking place within these vital organs.

The ureters, bladder and urethra are also key components of the renal system. Together, these components form a dynamic group that facilitates the storage and expulsion of urine, contributing to the body's fluid and electrolyte balance.

The renal system is responsible for a number of regulatory functions. From the delicate regulation of blood pressure to the adjustment of acid–base balance, the renal system ensures that the internal environment remains finely tuned for optimal cellular function.

RENAL SYSTEM

The renal system, also known as the urinary system, consists of:

- Two kidneys, which filter the blood to produce urine.
- Two ureters, which convey urine to the bladder.
- One urinary bladder, a storage organ for urine until it is eliminated.
- One urethra, which conveys urine to the exterior.

The organs of the renal system are depicted in Figure 1.1.

KIDNEYS (EXTERNAL)

There are usually two kidneys, one on each side of the spinal column (located in the posterior abdomen, retroperitoneally). They are approximately 11 cm long, 5–6 cm wide and 3–4 cm thick. The adrenal glands sit immediately superior to the kidneys. The kidneys are said to be bean-shaped organs where the outer border is convex; the inner border is known as the hilum (or hilus) and at this point, the renal arteries, renal veins, nerves and ureters enter and leave the kidneys (Figure 1.1). The renal artery transports blood to the kidneys and once the blood

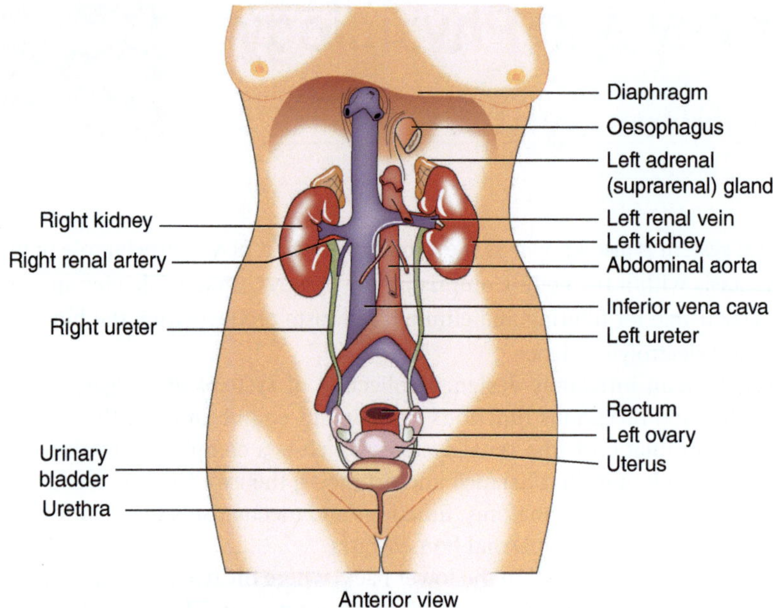

FIGURE 1.1   The renal system and female abdominal cavity

has been filtered, the renal vein takes the blood away. The right kidney is in contact with the liver's large right lobe and, as such, the right kidney is approximately 2–4 cm lower than the left kidney. There are three layers that cover and support the kidneys:

- Renal fascia

- Adipose tissue

- Renal capsule

The renal fascia provides the outer layer and consists of a thin layer of connective tissue that anchors the kidneys to the abdominal wall and the surrounding tissues. The middle layer is called the adipose tissue and surrounds the capsule. It cushions the kidneys from trauma. The inner layer is called the renal capsule. It consists of a layer of smooth connective tissue that is continuous with the outer layer of the ureter. The renal capsule protects the kidneys from trauma and maintains their shape (see Figure 1.2).

KIDNEYS (INTERNAL)

Inside the kidney, there are three distinct regions. They are:

- Renal cortex

- Renal medulla

- Renal pelvis

The outermost part of the kidney is the renal cortex. The renal cortex forms a continuous, smooth outer portion of the kidney with a number of projections (the renal columns) extending down between the pyramids. The renal column is the medullary extension of the renal cortex; it is reddish in colour and has a granular appearance, which is due to the

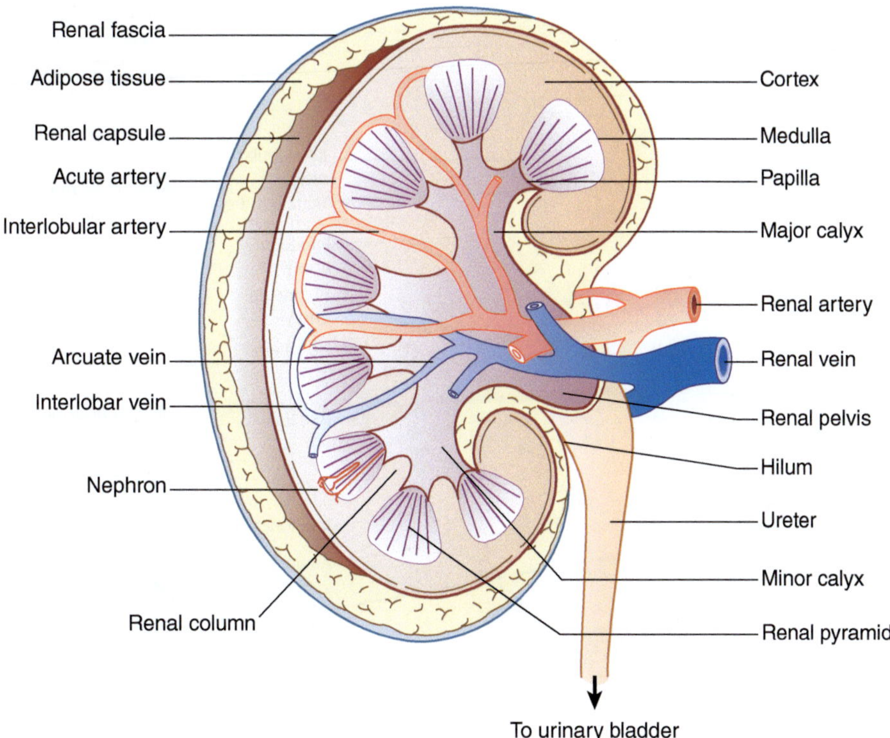

FIGURE 1.2 The external layers of the kidney

capillaries and the structures of the nephron. The medulla is lighter in colour with an abundance of blood vessels and tubules of the nephrons. The medulla is made up of approximately 8–12 renal pyramids. Figure 1.3a shows the frontal section of the right kidney and Figure 1.3b the path of blood flow. The renal pyramids, also called malpighian pyramids, are the cone-shaped sections of the kidneys. The wider portion of the cone faces the renal cortex, with the narrow ends pointing internally, and this section is known as the renal papilla. Urine formed by the nephrons flows into cup-like structures, called calyces, via papillary ducts. Each kidney contains approximately 8–18 minor calyces and two or three major calyces. The minor calyces receive urine from the renal papilla, which conveys the urine to the major calyces. The major calyces unite to form the renal pelvis, which then conveys urine to the bladder (see Figure 1.4). The renal pelvis forms the expanded upper aspect of the ureter, which is funnel-shaped, and it is in the region where two or three calyces converge.

NEPHRONS

The nephrons are small structures that form the functional units of the kidney. The nephron consists of a glomerulus and a renal tubule (see Figure 1.5). There are around one million nephrons in each kidney. It is within these structures that urine is formed. The nephrons:

- Filter blood

- Perform selective reabsorption

- Excrete unwanted waste products from the filtered blood

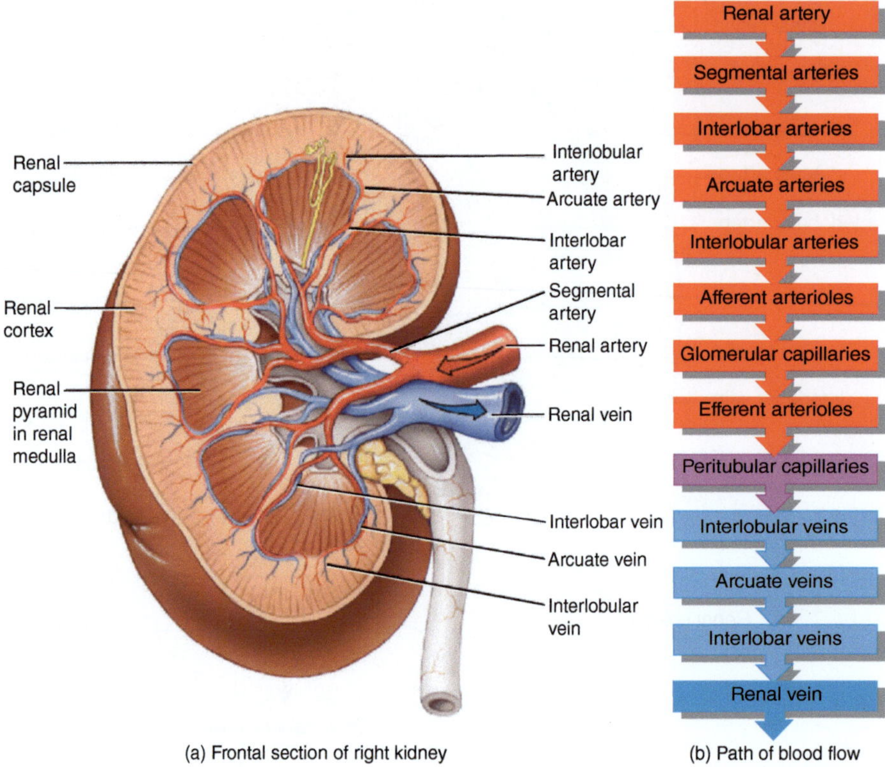

FIGURE 1.3 (a and b) The internal structures. *Source:* Tortora and Derrickson (2009). Reproduced with permission from John Wiley & Sons.

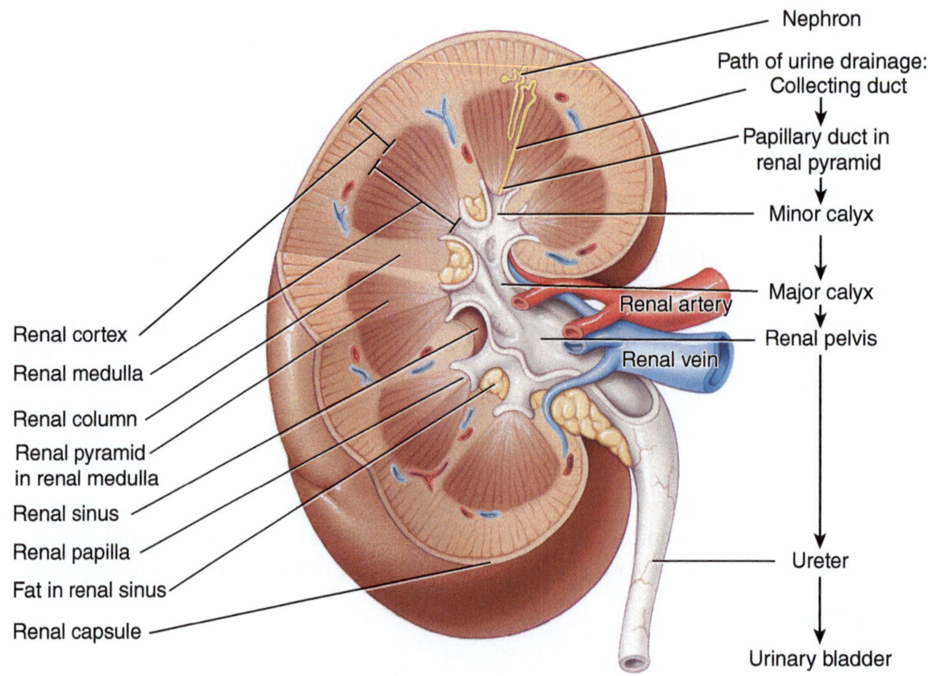

FIGURE 1.4 The internal structures showing blood vessels. *Source:* Tortora and Derrickson (2009). Reproduced with permission from John Wiley & Sons.

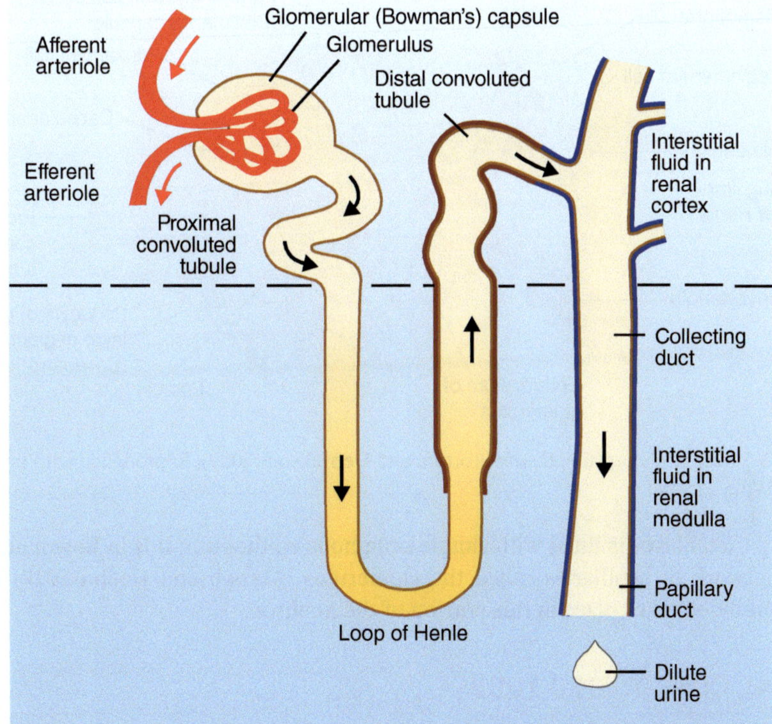

FIGURE 1.5   Nephron. *Source:* Tortora and Derrickson (2009). Reproduced with permission from John Wiley & Sons.

The nephron plays a key part in homeostasis. This system helps to regulate the amount of water, salts, glucose, urea and other minerals in the body. The nephron is a filtration system situated in the kidney and is responsible for the reabsorption of water and salts. The nephron is divided into several sections:

- Bowman's capsule

- Proximal convoluted tubule

- Loop of Henle

- Distal convoluted tubule (DCT)

- Collecting ducts

Each section carries out a different function; these are discussed in the following sections.

BOWMAN'S CAPSULE

Also known as the glomerular capsule (see Figure 1.6), Bowman's capsule is a cup-like sac and is the first portion of the nephron. Bowman's capsule is part of the filtration system in the kidneys. When blood reaches the kidneys for filtration, it first enters Bowman's capsule, with the capsule separating the blood into two components: a filtrated blood product and a filtrate that is moved through the nephron, another structure in the kidneys. The glomerular capsule consists of visceral and parietal layers. Epithelial cells, known as podocytes, line the visceral layer,

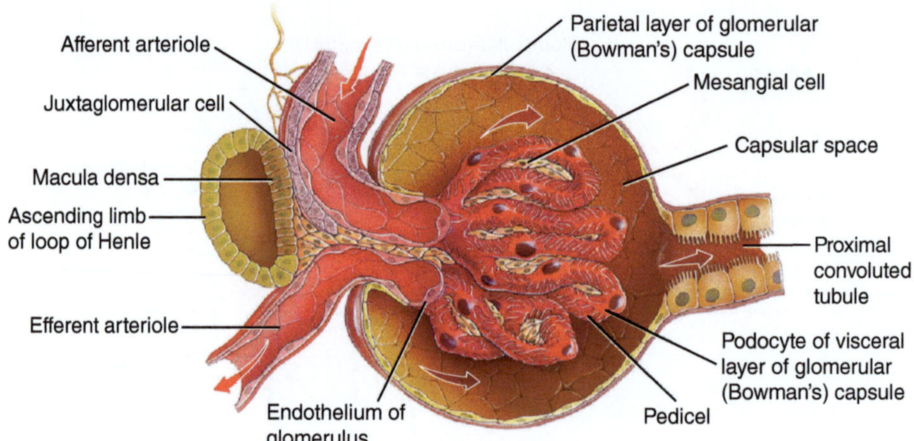

FIGURE 1.6   Bowman's capsule. *Source:* Tortora and Derrickson (2009). Reproduced with permission from John Wiley & Sons.

while the parietal layer is lined with simple squamous epithelium; it is in Bowman's capsule that the network of capillaries called the glomerulus (Marieb and Hoehn 2019) is found. Filtration of blood takes place in this portion of the nephron.

PROXIMAL CONVOLUTED TUBULE

From the Bowman's capsule, the filtrate drains into the proximal convoluted tubule (see Figure 1.6). The surface of the epithelial cells of this segment of the nephron is covered with densely packed microvilli, which increase the surface area of the cells, thus facilitating their resorptive function. The infolded membranes forming the microvilli are the site of numerous sodium pumps. Resorption of salt, water and glucose from the glomerular filtrate occurs in this section of the tubule; at the same time, certain substances, including uric acid and drug metabolites, are actively transferred from the blood capillaries into the tubule for excretion.

LOOP OF HENLE

The proximal convoluted tubule bends into a loop known as the loop of Henle (see Figure 1.5). The loop of Henle dips or 'loops' from the cortex into the medulla (descending limb) and then returns to the cortex (ascending limb). The ascending loop of Henle is much thicker than the descending portion. The main function of the loop of Henle is to generate a concentration gradient creating a region of a high concentration of sodium in the medulla of the kidney. The descending portion of the loop of Henle is highly permeable to water with low permeability to ions and urea. The ascending loop of Henle is permeable to ions but not to water. When required, urine is concentrated in this portion of the nephron. This is possible because of the high concentration of solute in the substance or interstitium of the medulla. Different sections of the loop of Henle have different actions:

- The descending loop of Henle is relatively impermeable to solute but permeable to water, water moves out by osmosis and the fluid in the tubule becomes hypertonic.

- The thin section of the ascending loop of Henle is virtually impermeable to water, but permeable to solute, particularly sodium and chloride ions. As sodium and chloride ions move out down the concentration gradient, fluid within the tubule first becomes isotonic and then hypotonic as more ions leave. Urea diffuses into the ascending limb, keeping the urea within the interstitium of the medulla, where it also has a role in concentrating urine.

- The thick section of the ascending loop of Henle and early distal tubule are virtually impermeable to water. However, sodium and chloride ions are actively transported out of the tubule, making the tubular fluid very hypotonic.

DISTAL CONVOLUTED TUBULE

The thick ascending portion of the loop of Henle leads into the DCT (see Figure 1.6). The DCT is lined with simple cuboidal cells, and the lumen of the DCT is larger than the proximal convoluted tubule lumen because the proximal convoluted tubule has a brush border (microvilli). The DCT is an important site:

- It actively secretes ions and acids.

- It plays a part in the regulation of calcium ions by excreting excess calcium ions in response to calcitonin (a hormone).

- It selectively reabsorbs water.

- Arginine vasopressin receptor 2 proteins are also located there.

- It plays a role in regulating pH by absorbing bicarbonate and secreting protons (H^+) into the filtrate.

The final concentration of urine, in this section, is dependent on antidiuretic hormone (ADH). If ADH is present, the distal tubule and the collecting duct become permeable to water. As the collecting duct passes through the medulla with a high solute concentration in the interstitium, the water moves out of the lumen of the duct and concentrated urine is formed. In the absence of ADH, the tubule is minimally permeable to water, so a large volume of dilute urine is formed.

COLLECTING DUCTS

The DCT then drains into the collecting ducts (see Figure 1.5). Several collecting ducts converge, draining into a larger system called the papillary ducts, which in turn empty into the minor calyx (plural: calices). From here, the filtrate, now called urine, drains into the renal pelvis. This is the final stage where sodium and water are reabsorbed. In dehydration, approximately 25% of the water filtered is reabsorbed in the collecting duct. The cells of the collecting ducts are impermeable to water. With the aid of the ADH and aquaporins, water is reabsorbed from the collecting ducts. Aquaporins are proteins embedded in the cell membrane that regulate the flow of water. Aquaporins selectively transport water molecules in and out of the cell, while preventing the passage of ions and other solutes. Aquaporin 1 is abundant in the proximal convoluted tubule and the descending thin limb of the loop of Henle, and aquaporins 2, 3 and 4 are present in the collecting ducts; however, aquaporin 4 is predominantly found in the brain.

THE FUNCTIONS OF THE KIDNEY

The kidneys maintain fluid balance, electrolyte balance and the acid–base balance of the blood, removing waste products and excess water (fluid) collected by, and carried in, the blood as it flows through the body. Approximately 190 L of blood enters the kidneys daily via the renal arteries. Millions of tiny filters, called glomeruli, inside the kidneys separate waste products and water from the blood. Most of these unwanted substances come from what is ingested. The kidneys automatically remove the right amount of salt and other minerals from the blood to leave just the quantities the body needs.

In removing just the right amount of excess fluid, healthy kidneys maintain the body's fluid balance. Fluid content stays at about 55% of total weight in women; in men, this is about 60% of total weight. The kidneys maintain these proportions by balancing the amount of fluid that leaves the body against the amount entering the body. When a large volume of fluid is drunk, healthy kidneys remove the excess fluid, producing a lot of urine. If fluid intake is low, the kidneys retain fluid and not much urine is passed. Fluid also leaves the body via sweat, breath and faeces. In hot weather, a lot of fluid is lost by sweating and the kidneys do not produce much urine.

Kidneys synthesise hormones such as renin and angiotensin. These hormones regulate how much sodium and fluid the body retains and how well the blood vessels can expand and contract. This, in turn, helps control blood pressure.

Kidneys produce a hormone – erythropoietin, which is transported in the blood to the bone marrow here it stimulates the production of red blood cells. These cells carry oxygen throughout the body. Without enough healthy red blood cells anaemia develops.

Healthy kidneys keep bones strong by producing calcitriol. Calcitriol maintains the right levels of calcium and phosphate in the blood and bones. Calcium and phosphate balance is important in keeping bones healthy. When the kidneys fail, they may not produce enough calcitriol, leading to abnormal levels of phosphate, calcium and vitamin D, resulting in renal bone disease. A summary of the functions of the kidney is found in Box 1.1.

<table>
<tr><td>BOX 1.1</td><td>FUNCTIONS OF THE KIDNEY</td></tr>
</table>

- Regulation of electrolytes: regulates ions such as sodium, potassium, calcium, chloride and phosphate.
- Regulation of blood pH: excretes hydrogen ions into urine and conserves bicarbonate ions, thus helping regulate the pH of blood.
- Regulation of blood volume: by conserving or eliminating water in urine.
- Secretes renin (regulates blood pressure) and erythropoietin (production of red blood cells).
- Production of calcitriol for regulation of calcium levels.
- Aids in the regulation of blood glucose level by gluconeogenesis.
- Detoxification of free radicals and drugs.
- Excretion of waste products, such as urea, uric acid and creatinine.

Source: Nagalingam (2020).

KIDNEY BLOOD SUPPLY

The kidney filters at least 20–25% of blood during the resting cardiac output. Approximately 1200 mL of blood flows through the kidney each minute. Both kidneys receive their blood supply directly from the aorta via the renal artery, which is divided into anterior and posterior renal arteries. There are several arteries delivering blood to the kidneys:

- Renal artery: arises from the abdominal aorta at the level of the first lumbar vertebra.

- Segmental artery: branch of the renal artery.

- Interlobar artery: branch of the segmental artery.

- Arcuate artery: renal columns leading to the corticomedullary junction.

- Interlobular arteries: divisions of the arcuate arteries.

The branches of the interlobular artery enter the nephrons as afferent arterioles. Each nephron receives one afferent arteriole, further subdivided into a tuft of capillaries called the glomerulus. The glomerular capillaries reunite, leaving Bowman's capsule as efferent arterioles. Efferent arterioles unite, forming peritubular capillaries and then interlobular veins that unite to form the arcuate veins and finally the interlobar veins. Blood leaves the kidneys through the renal vein, which then flows into the inferior vena cava. The diameter of the afferent arteriole is larger than the diameter of the efferent arteriole.

URINE FORMATION

There are three processes involved in the formation of urine:

- Filtration

- Selective reabsorption

- Secretion

FILTRATION

Urine formation starts with the process of filtration, which goes on continually in the renal corpuscles. Filtration takes place in the glomerulus within Bowman's capsule. The blood for filtration is supplied by the renal artery. In the kidney, the renal artery divides into smaller arterioles. The arteriole entering Bowman's capsule is the afferent arteriole, further subdividing into a cluster of capillaries called the glomerulus.

As blood passes through the glomeruli, much of its fluid, containing both useful chemicals and dissolved waste materials, soaks out of the blood through the membranes (by osmosis and diffusion). Here it is filtered and flows into Bowman's capsule. This process is called glomerular filtration. The water, waste products, salt, glucose and other chemicals that have been filtered out of the blood are collectively known as glomerular filtrate.

The fluid from the filtered blood is protein-free but contains electrolytes such as sodium chloride, potassium and waste products of cellular metabolism, for example, urea, uric acid and creatinine (McCance and Huether 2018). The filtered blood returns into circulation via the efferent arteriole and finally into the renal vein.

SELECTIVE REABSORPTION

Selective reabsorption processes ensure that any substances in the filtrate essential for body function are reabsorbed into the plasma. Substances, for example, sodium, calcium, potassium and chloride, are reabsorbed to maintain fluid and electrolyte balance and the pH of the blood. However, if these substances are in excess of body requirements, they are excreted in the urine. Only 1% of the glomerular filtrate actually leaves the body; 99% is reabsorbed into the bloodstream (Swales 2022). The reabsorption occurs via three processes:

- Osmosis

- Diffusion

- Active transport

Blood glucose is entirely reabsorbed into the blood from the proximal tubules. It is actively transported out of the tubules and into the peritubular capillary blood. None of these valuable nutrients are wasted by being lost in the urine. Sodium (Na^+) and other ions are only partially reabsorbed from the renal tubules into the blood. For the most part, however, sodium ions are actively transported back into the blood from the tubular fluid. The amount of sodium reabsorbed varies, depending largely on how much salt is taken in from the food eaten.

As a person increases their salt intake, the kidneys decrease the amount of sodium reabsorption into the blood. That is, more sodium is retained in the tubules. Therefore, the amount of salt excreted in the urine increases. The process also works the other way. The less the salt intake, the greater the amount of sodium reabsorbed into the blood, and the amount of salt excreted in the urine decreases.

EXCRETION

Substances not removed through filtration are secreted into the renal tubules from the peritubular capillaries (see Figure 1.7) of the nephron (Martini, Nath, and Bartholomew 2018). These include drugs and hydrogen ions. Tubular secretion primarily takes place by active transport, a process by which substances are moved across biological membranes. Tubular secretion occurs from epithelial cells lining the renal tubules and the collecting ducts. Substances secreted into the tubular fluid include:

- Potassium ions (K^+)

- Hydrogen ions (H^+)

- Ammonium ions (NH^{4+})

- Creatinine

- Urea

- Some hormones

It is the tubular secretion of hydrogen and ammonium ions that helps to maintain the pH of the blood.

HORMONAL CONTROL OF TUBULAR REABSORPTION AND SECRETION

Four hormones play a role in the regulation of fluid and electrolytes:

- Angiotensin II

- Aldosterone

- ADH

- Atrial natriuretic peptide (ANP)

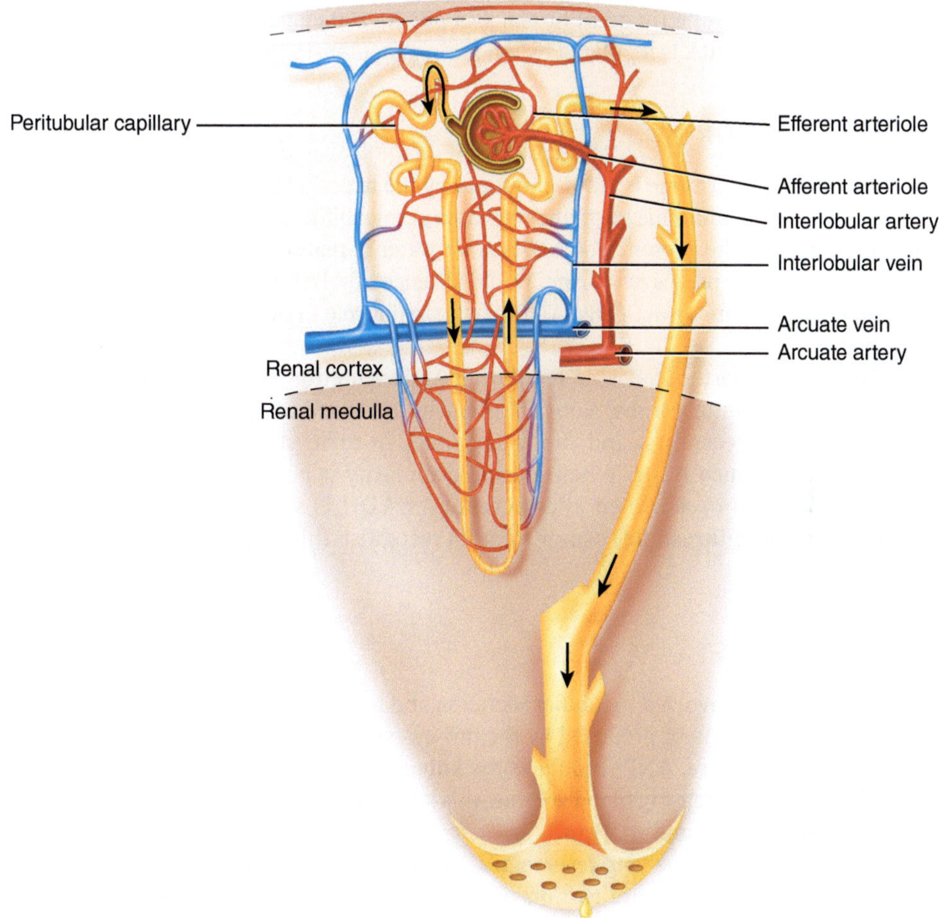

FIGURE 1.7 Nephron with capillaries. *Source:* Tortora and Derrickson (2009). Reproduced with permission from John Wiley & Sons.

ANGIOTENSIN AND ALDOSTERONE

As blood volume and blood pressure decrease, the juxtaglomerular cells secrete a hormone called renin. Juxtaglomerular cells are found near the glomerulus. These cells synthesise, store and secrete renin. Renin acts on a plasma protein called angiotensinogen and converts it into angiotensin I. Angiotensinogen is produced by the hepatocytes. Angiotensin I is transported by the blood to the lungs. In the lung capillaries, there are enzymes called angiotensin-converting enzymes (ACEs). ACE is predominantly located in lung capillaries, but is also found throughout the body. ACE converts angiotensin I into angiotensin II. Angiotensin II is a short-acting, powerful vasoconstrictor, thus increasing blood pressure (Jones and Stegall 2019). Angiotensin II promotes the reabsorption of sodium, chloride and water in the proximal convoluted tubule. It also has an effect on the release of aldosterone.

Aldosterone, a steroid hormone, is secreted by the adrenal glands. It serves as the principal regulator of the salt and water balance of the body and is categorised as a mineralocorticoid. It also has a small effect on the metabolism of fats, carbohydrates and proteins.

Aldosterone is synthesised in the body from corticosterone, a steroid derived from cholesterol. Production of aldosterone (in adult humans, about 20–200 μg per day) in the zona glomerulosa of the adrenal cortex is regulated by the renin–angiotensin system.

ANTIDIURETIC HORMONE

The third principal hormone is ADH, produced by the hypothalamus gland and stored by the posterior pituitary gland. This hormone increases permeability of the cells in the DCT and the collecting ducts. In the presence of ADH, more water is reabsorbed from the renal tubules; therefore, the patient passes less urine. In the absence of ADH, less water is reabsorbed, and the patient passes more urine. Thus, ADH plays a major role in the regulation of fluid balance in the body.

The most important variable regulating ADH secretion is plasma osmolarity, or the concentration of solutes in the blood. Osmolarity is sensed in the hypothalamus by neurones called osmoreceptors, and those neurones, in turn, stimulate secretion from the neurones that produce ADH. When plasma osmolarity is below a certain threshold, the osmoreceptors are not activated and secretion of ADH is suppressed. When osmolarity increases above the threshold, osmoreceptors recognise this and stimulate the neurones that secrete ADH.

ATRIAL NATRIURETIC PEPTIDE

The fourth hormone involved in tubular secretion and reabsorption is ANP. ANP is a powerful vasodilator and is a protein produced by the myocytes of the atria of the heart in response to increased blood pressure. ANP stimulates the kidneys to excrete sodium and water from the renal tubules, thus decreasing blood volume; this lowers blood pressure. ANP also inhibits the secretion of aldosterone and ADH.

ANP is involved in the long-term regulation of sodium and water balance, blood volume and arterial pressure. There are two major pathways of natriuretic peptide actions: vasodilator effects and renal effects, which lead to natriuresis and diuresis. ANP directly dilates veins (increases venous compliance) and thereby decreases central venous pressure, reducing cardiac output by decreasing ventricular preload. ANP also dilates arteries, which decreases systemic vascular resistance and systemic arterial pressure.

COMPOSITION OF URINE

Urine is normally a sterile and clear fluid of nitrogenous waste and salts. It is translucent with an amber or light yellow colour. Its colour is due to the pigments from the breakdown of haemoglobin. Concentrated urine tends to be darker in colour than normal urine. However, other factors, for example, diet, medications and certain diseases, may affect the colour of urine. It is slightly acidic and the pH ranges from 4.5 to 8. The pH is affected by an individual's dietary intake and state of health. A diet high in animal protein tends to make the urine more acidic, while a vegetarian diet may make the urine more alkaline. The volume of urine produced depends on the circulating volume of blood. ADH regulates the amount of urine passed by the individual. If the person is dehydrated, more ADH is released from the posterior pituitary gland, resulting in water reabsorption and less urine being produced. However, if the person has consumed a large amount of fluid, which increases the circulating volume, less ADH is released and more water is passed as urine.

Urine is 96% water and approximately 4% solutes derived from cellular metabolism. The solutes include organic and inorganic waste products and unwanted substances such as drugs. Normally, there is no protein or blood present in the urine; if these are present, then the person may have a medical condition.

CHARACTERISTICS OF NORMAL URINE

The volume produced is one of the physical characteristics of urine. Other physical characteristics that can apply to urine include colour, turbidity (transparency), smell (odour), pH (acidity/alkalinity) and density.

Colour: It is usually yellow–amber, but varies according to recent diet, medication and the concentration of the urine. Drinking more water generally reduces the concentration of urine and therefore causes it to have a lighter colour. However, if a person does not drink a large amount of fluid, this may increase the concentration and the urine will have a darker colour.

Smell: The smell, or odour, of urine may provide health information. For example, the urine of diabetics may have a sweet or fruity odour due to the presence of ketones (organic molecules of a particular structure). Generally, fresh urine has a mild smell, but stale urine or infected urine has a stronger odour, similar to ammonia. Cloudy urine may indicate infection, whereas foamy urine can indicate the presence of protein and glucose.

Acidity: pH is a measure of the acidity (or alkalinity) of a solution. The pH of a substance (solution) is usually represented as a number in the range from 0 (strong acid) to 14 (strong alkali, also known as a 'base'). Pure water is 'neutral', neither acid nor alkali; it has a pH of 7. The pH of normal urine is generally in the range of 4.5–8, with a typical average being around 6.0.

Specific gravity: Specific gravity is also known as 'relative density'. This is the ratio of the weight of a volume of a substance compared to the weight of the same volume of distilled water. Given urine is mostly water, but it also contains some other substances dissolved in the water, its relative density is expected to be close to, but slightly greater than, 1.000.

URETERS

The ureters are tubular organs running from the renal pelvis to the posterolateral base of the urinary bladder. They are approximately 25–30 cm in length and 5 mm in diameter (Martini, Nath, and Bartholomew 2018). The ureters terminate at the bladder and enter obliquely through the muscle wall of the bladder. They pass over the pelvic brim at the bifurcation of the common iliac arteries (see Figure 1.8).

The ureters have three layers:

- Transitional epithelial mucosa (inner layer)

- Smooth muscle layer (middle layer)

- Fibrous connective tissue (outer layer)

Urine is transported through the ureters via muscular movements of the urinary tract's peristaltic muscular waves. When the renal pelvis becomes laden with urine, the peristaltic wave action encourages urine to leave the pelvis. The amount of urine in the renal pelvis determines the frequency of the peristaltic wave action, which can range from one wave every few minutes to one wave every few seconds. This action creates a pressure force that moves the urine through the ureters and into the bladder in small spurts.

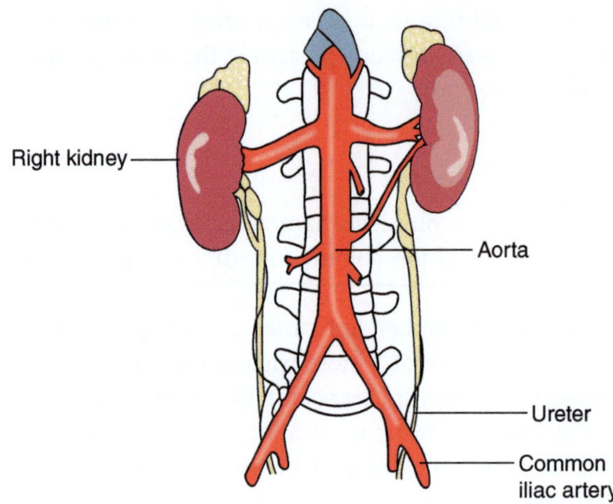

FIGURE 1.8 Common iliac vessels and ureter. *Source:* Nair and Peate (2009). Reproduced with permission from John Wiley & Sons.

URINARY BLADDER

The urinary bladder is a hollow muscular organ, located in the pelvic cavity posterior to the symphysis pubis. In the male, the bladder lies anterior to the rectum; in the female, it lies anterior to the vagina and inferior to the uterus (Martini, Nath, and Bartholomew); it is a smooth muscular sac storing urine. Although the shape of the bladder is spherical, this is altered by the pressure of surrounding organs. When the bladder is empty, the inner section of the bladder forms folds (rugae). As the bladder fills with urine the walls of the bladder become smoother. As urine accumulates, the bladder expands without a significant rise in the internal pressure of the bladder. The bladder normally distends and holds approximately 350–750 mL of urine. In females, the bladder is slightly smaller, as the uterus occupies the space above the bladder.

The inner lining of the urinary bladder is a mucous membrane of transitional epithelium, continuous with that in the ureters. When the bladder is empty, the rugae and transitional epithelium allow the bladder to expand as it fills. The second layer on the walls is the submucosa, which supports the mucous membrane. It is composed of connective tissue with elastic fibres.

The inner floor of the bladder includes a triangular section called the trigone. The trigone is formed by three openings in the floor of the urinary bladder. Two of the openings are from the ureters and form the base of the trigone. Small flaps of mucosa cover these openings, acting as valves allowing urine to enter the bladder but preventing it from backing up from the bladder into the ureters. The third opening, at the apex of the trigone, is the opening into the urethra (see Figure 1.9). A band of the detrusor muscle encircles this opening to form the internal urethral sphincter.

The walls of the bladder consist of muscle fibres:

- Transitional epithelial mucosa

- A thick muscular layer

- A fibrous outer layer

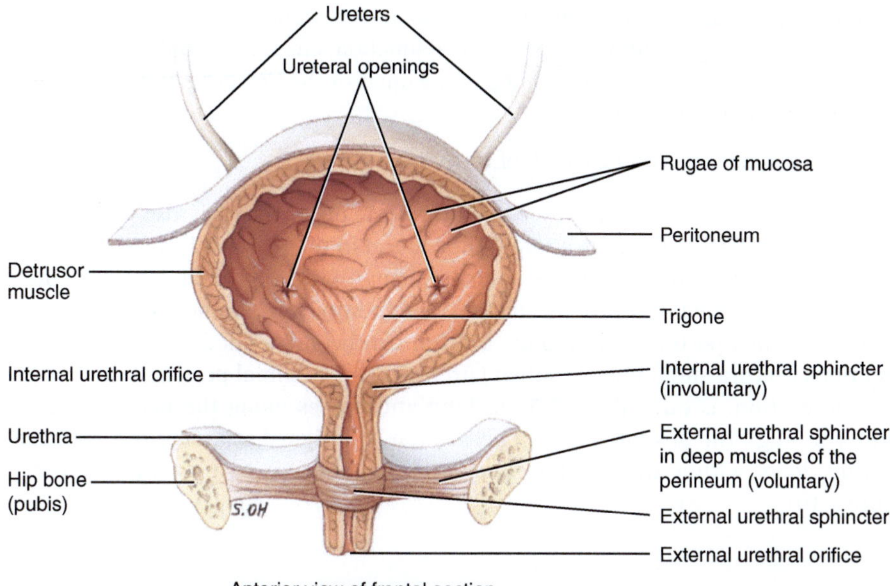

FIGURE 1.9 Layers of the urinary bladder. *Source:* Tortora and Derrickson (2009). Reproduced with permission from John Wiley & Sons.

The urinary tract can become blocked or obstructed (e.g. from a kidney stone, tumour, expanding uterus during pregnancy or enlarged prostate gland). The build-up of urine can lead to infection and injury to the kidney (Peate 2020).

Urinary tract infections, such as cystitis (an infection of the bladder), can lead to more serious infections further up the urinary tract. An obstruction in the urinary tract can make a kidney infection more likely. Infections elsewhere in the body, including, for example, streptococcal infections, the skin infection impetigo or a bacterial infection in the heart, can also be carried through the bloodstream to the kidney and cause a problem there.

URETHRA

The urethra is a muscular tube draining urine from the bladder, conveying it out of the body. It contains three coats: muscular, erectile and mucous; the muscular is the continuation of the bladder muscle layer. The urethra is encompassed by two separate urethral sphincter muscles. The internal urethral sphincter muscle is formed by involuntary smooth muscles; the lower voluntary muscles make up the external sphincter muscles. The internal sphincter is created by the detrusor muscle. The urethra is longer in males than in females. Sphincters keep the urethra closed when urine is not being passed. The internal urethral sphincter is under involuntary control and lies at the bladder–urethra junction. The external urethral sphincter is under voluntary control.

MALE URETHRA

The male urethra passes through four different regions:

1. Prostatic region: passes through the prostate gland.

2. Membranous portion: passes through the pelvic diaphragm.

3. Bulbar urethra: located inside the perineum and scrotum, extends from the external distal urinary sphincter to the penoscrotal junction, and is surrounded by the corpus spongiosum. It contains the opening of the ducts of the Cowper glands and differs in length from person to person.

4. Penile region: extends the length of the penis.

In the male, the urethra not only excretes fluid waste products but is also part of the reproductive system. Rather than the straight tube found in the female's body, the male urethra is S-shaped to follow the line of the penis. It is approximately 20 cm long. The male urethra can be segregated into various portions: the spongy portion, the prostatic portion and the membranous portion. The spongy urethra can be subdivided into fossa navicularis, pendulous urethra and bulbous (bulbar) urethra. The proximal portion, which is also the prostatic portion, is only about 2.5 cm long and passes along the neck of the urinary bladder through the prostate gland. This section is designed to accept the drainage from the tiny ducts within the prostate and is equipped with two ejaculatory tubes (see Figure 1.10).

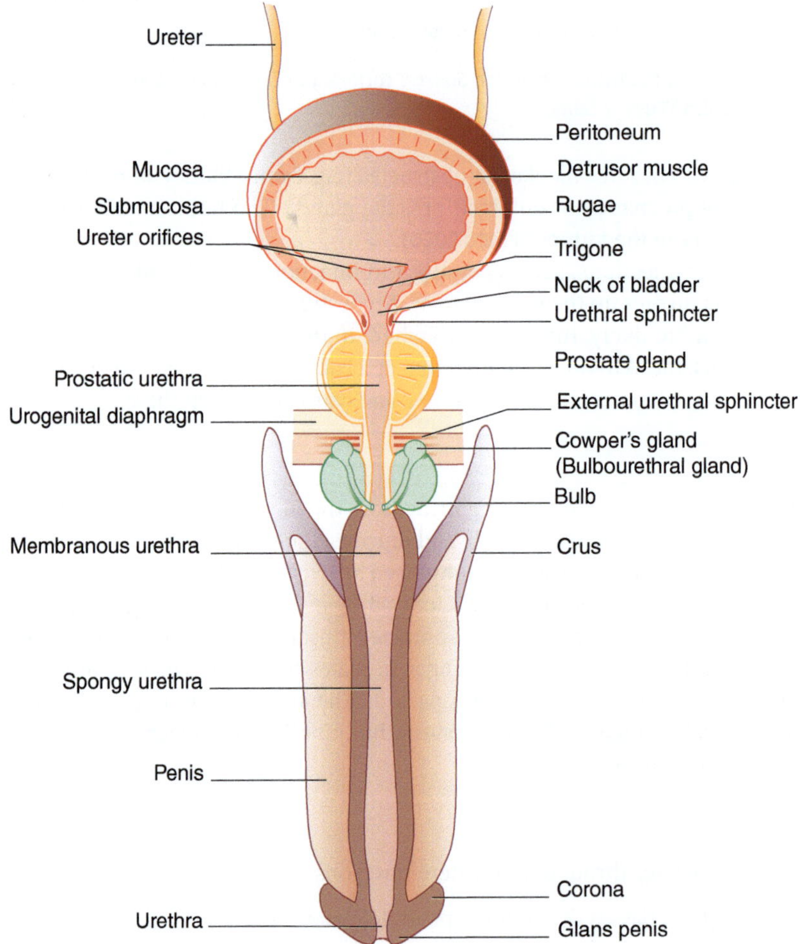

FIGURE 1.10　Male urethra

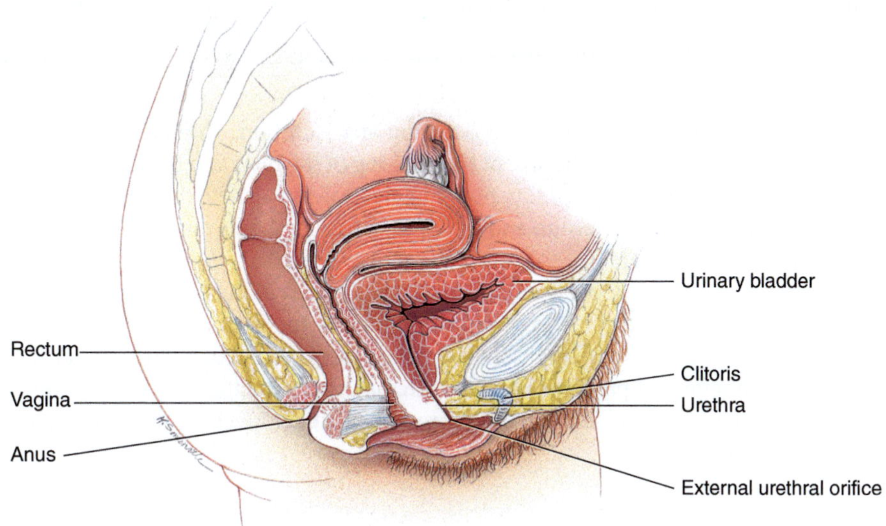

FIGURE 1.11 Location of the female urethra. *Source:* Nair and Peate (2013). Reproduced with permission from John Wiley & Sons.

FEMALE URETHRA

The female urethra is bound to the anterior vaginal wall. The external opening of the urethra is anterior to the vagina and posterior to the clitoris. The urethra in the female is approximately 4 cm long and leads out of the body via the urethral orifice. In the female, the urethral orifice is located in the vestibule of the labia minora, between the clitoris and the vaginal orifice. In the female body, the urethra's only function is to transport urine out of the body (see Figure 1.11).

MICTURITION

When the volume of urine in the bladder reaches approximately 300 mL, stretch receptors in the bladder walls are stimulated and excite sensory parasympathetic fibres relaying information to the sacral area of the spine. This information is assimilated into the spine and relayed to two different sets of neurones. Parasympathetic motor neurones are excited and act to contract the detrusor muscles in the bladder so that bladder pressure increases and the internal sphincter opens. At the same time, somatic motor neurones supplying the external sphincter via the pudendal nerve are inhibited, allowing the external sphincter to open and urine to flow out, assisted by gravity.

A person usually has control over bladder function. They can increase or decrease the rate of flow of urine and stop and start at will (unless there are physiological problems), thus making micturition a simple reflex action.

CONCLUSION

The renal system consists of the kidneys, ureters, urinary bladder and the urethra. Collectively, these components play an important role in maintaining homeostasis. The renal system removes waste products of metabolism, secretes hormones, regulates fluid balance and maintains homeostasis. Some of the functions it carries out include:

- Regulating blood volume through urine production and blood pressure by releasing renin.

- Regulating electrolyte balance in the body through hormones such as aldosterone.

- Maintaining acid–base balance by regulating the secretion of hydrogen and bicarbonate ions.

- Excreting waste products (e.g. urea and uric acid) and conserving valuable nutrients essential for the body.

Urine is formed by filtration, selective reabsorption and secretion. The selectivity of the glomerular filtrate is determined by the size of the opening of the filter and blood pressure. There are other factors regulating urine production and electrolyte balance; they include hormone regulation such as ADH, aldosterone and ANP hormones and neuronal regulation through the autonomic nervous system.

The urinary bladder is a storage organ for urine, located in the pelvic cavity. It contains three layers: the muscular, erectile and mucous layers. Urine is stored in the bladder until the person gets the urge to empty their bladder. The process of micturition is under the control of the sympathetic and parasympathetic system. During micturition, strong muscles in the bladder walls (the detrusor muscles) compress the bladder, pushing its contents into the urethra, thus voiding urine.

GLOSSARY OF TERMS

Anterior: Referring to the front.

Bifurcation: Dividing into two branches.

Calyces: Small, funnel-shaped cavities formed from the renal pelvis.

Diuresis: Excessive urine production.

Erythropoietin: Hormone produced by the kidneys that regulates red blood cell production.

Excretion: The elimination of waste products from metabolism.

Filtration: A passive transport system.

Glomerulus: A network of capillaries found in Bowman's capsule.

Hilum (hilus): An indention near the centre of the concave area of the kidney, where its vessels, nerves and ureter enter/leave.

Nephron: Functional unit of the kidney.

Posterior: Pertaining to the back, behind.

Renal artery: A blood vessel that takes blood to the kidney.

Renal cortex: The outermost part of the kidney.

Renal medulla: The middle aspect of the kidney.

Renal pelvis: The funnel-shaped section of the kidney.

Renal pyramids: Cone-shaped structures of the medulla.

Renal vein: A blood vessel that returns filtered blood into circulation.

Renin: A renal hormone that alters systemic blood pressure.

Sphincter: A ring-like muscle fibre that can constrict.

Ureter: Membranous tube that drains urine from the kidneys to the bladder.

Urethra: Muscular tube that drains urine from the bladder.

MULTIPLE CHOICE QUESTIONS

1. What is the primary function of the renal system?
    a) Digestion
    b) Filtration of blood
    c) Respiratory exchange
    d) Hormone secretion

2. Where does filtration of blood take place in the kidney?
    a) Glomerulus
    b) Renal cortex
    c) Renal medulla
    d) Ureter

3. Which hormone stimulates the production of red blood cells in the bone marrow?
    a) Insulin
    b) Thyroxine
    c) Erythropoietin
    d) Adrenaline

4. What is the normal function of the loop of Henle in the nephron?
    a) Filtration
    b) Reabsorption of water and electrolytes
    c) Secretion of hormones
    d) Storage of urine

5. What is the purpose of the ureters in the renal system?
    a) Filtration of blood
    b) Storage of urine
    c) Transport of urine from the kidneys to the bladder
    d) Reabsorption of water

6. The bladder is a muscular organ that primarily functions for:
    a) Filter blood
    b) Store urine
    c) Produce hormones
    d) Reabsorb water

7. The functional unit of the kidney responsible for filtration, reabsorption and secretion is called:
    a) Nephron
    b) Glomerulus
    c) Renin
    d) Ureter

8. What is the role of angiotensin-converting enzyme (ACE) in the renal system?
    a) Stimulating erythropoiesis
    b) Regulating blood pressure
    c) Facilitating urine excretion
    d) Enhancing glucose metabolism

9. What is the primary function of the nephron's distal convoluted tubule?
   a) Filtration
   b) Reabsorption of water
   c) Secretion of substances into urine
   d) Initial filtration of blood

10. What hormone acts on the distal tubules and collecting ducts to promote the reabsorption of sodium and water?
    a) Aldosterone
    b) Antidiuretic hormone (ADH)
    c) Renin
    d) Erythropoietin

REFERENCES

Jones, K. and Stegall, M. (2019). Nursing patients with urinary disorders (Chapter 9). In: *Alexander's Nursing Practice*, 5e. (ed. I. Peate). London: Elsevier.

Marieb, E. N. and Hoehn, K. (2019). *Human Anatomy and Physiology*, 11e. Harlow: Pearson Education.

Martini, F. H., Nath, J. L., and Bartholomew, E. F. (2018). *Fundamentals of Anatomy and Physiology*, 11e, Global Edition. Harlow: Pearson Education.

McCance, K. L. and Huether, S. E. (2018). *Pathophysiology: The Biologic Basis for Disease in Adults and Children*, 8e. St Louis: Mosby.

Nagalingam, K. (2020). The renal system (Chapter 11). In: *Fundamentals of Anatomy and Physiology*, 3e. (ed. I. Peate and S. Evans). Oxford: Wiley.

Peate, I. (2020). *Anatomy and Physiology for Nursing and Healthcare Students at a Glance*, 2e. Oxford: Wiley.

Swales. (2022). The person with a urinary disorder (Chapter 29). In: *Nursing Practice*, 3e. (ed. I. Peate and A. Mitchell). Oxford: Wiley.

The renal system is made up of a number of components, which play a key role in maintaining homeostasis. It is the responsibility of the kidneys to filter the blood, remove waste products and produce urine. The formation of urine is achieved through the processes of filtration, selective reabsorption and excretion. The kidneys also have an endocrine function, secreting hormones such as renin and erythropoietin. The bladder acts as a reservoir for the urine. As it fills, its shape becomes ovoid and rises out of the pelvis in the midline towards the umbilicus, behind the anterior abdominal wall. The bladder wall contains a layer of smooth muscle, the detrusor, which contracts under parasympathetic control, allowing urine to pass through the urethra (micturition). The conscious desire to micturate occurs when the bladder holds approximately 250–350 mL of urine (Dhauan and Kluth 2024).

Having an understanding of the various roles and functions of the renal system, including the component parts, can help in performing a holistic assessment to provide safe and effective care.

Assessing the renal system allows healthcare professionals to evaluate kidney function, detect abnormalities and monitor the effectiveness of interventions. This chapter provides an overview of the key components involved in assessing the renal system, encompassing both subjective and objective measures.

ASSESSING NEEDS

The renal system is essential to the function of the entire body. By working with the other systems of the body, the renal system helps to maintain homeostasis. Undertaking an accurate assessment can help to ensure that any impact on the renal system is recognised quickly and steps are taken to avoid damage to the kidneys or other components of this system. The assessment of the renal system is a fundamental aspect of care provision, enabling healthcare professionals to gain insights into the status of renal function and identify potential issues. Treating renal failure early can make a difference to the health and well-being of the individual and their family. Having an understanding of those people who are at risk will enable prompt and purposeful targeting of those in need.

As with the assessment of all body systems, there are two types of data that can be collected:

- Subjective data includes reports from the patient (primary data or primary source) and data from other sources (secondary data).

- Objective data includes data derived from the physical examination and any diagnostic tests.

When the data is taken as a whole, considered together, it provides an understanding of how the renal system is functioning (or not). It should be noted that the absence of signs or symptoms does not always mean that the system is functioning normally.

HISTORY TAKING

The focus of history taking involves not only identifying signs and symptoms of illness but also the individual's experience of illness. History taking in a healthcare setting refers to the systematic process of collecting relevant information about a patient's medical history, current symptoms, lifestyle and other pertinent details. It is a critical component of the healthcare provider's initial assessment and ongoing patient care.

History taking is a central element of patient assessment, facilitating the delivery of high-quality care. Understanding the complexity and the processes involved in history taking allows healthcare providers to develop a better understanding of patients' problems. Care priorities can be recognised, and the most appropriate interventions put in place to optimise patient outcomes. The information obtained during history taking helps healthcare professionals make accurate diagnoses, develop appropriate treatment plans and understand the patient's individual needs (Fawcett and Rhynas 2012). The process typically involves the following components:

- Chief complaint

- Present health status

- Past health history

- Current lifestyle

- Psychosocial status

- Family history

- Physical assessment

Skilled history taking, according to Diamond-Fox (2023), has the potential to make a significant contribution to several fundamental care outcomes: patient satisfaction, patient concordance with prescribed therapies/interventions, overall diagnostic accuracy and positive patient outcomes.

The overall aim of the healthcare assessment is to identify symptoms and physical manifestations that bring together a wide range of pathologies that may be highly suggestive or even diagnostic of one such pathology or multiple concurrent pathologies.

The core of clinical practice often revolves around the interaction between a patient and a care provider. In its most simplistic form, it serves as the avenue through which individuals, whether unwell or perceiving themselves to be so, seek guidance from a trusted clinician. Traditionally, this encounter takes place in person, yet the use of non-face-to-face methods, such as remote consultations via telephone, video technology or online platforms, is becoming more prevalent (Fairhurst, Innes, and Dover 2024).

EFFECTIVE COMMUNICATION

The medical history takes account of the uniqueness of patients and the social, cultural and psychological factors that may play a role in illness. Medical history taking should follow a structured format for gathering information (Dover, Fairhurst, and Innes 2024). Mnemonics (or mental grids or frameworks), (Diamond-Fox 2023) are often used to assist and promote comprehensive history taking (see Table 2.1). Frameworks should not be viewed as tick lists, as they are a means of triggering the person taking the history to address particular areas of importance.

At the centre of effective history taking is skilled and patient-centred communication through which a rapport is established. Patients and healthcare providers work together to achieve a shared understanding of the nature of the problem and the way to address it. The Calgary–Cambridge framework (Kurtz and Silverman 1996) is an approach that can be used to obtain an accurate and detailed patient history. The Calgary–Cambridge framework consists of a set of key components that guide the process of effective communication during a medical encounter. Stages of the Calgary–Cambridge framework are outlined in Table 2.2.

The Calgary–Cambridge framework emphasises the importance of patient-centred care, empathy and clear communication in the clinical setting. The framework provides a structured and comprehensive approach to communication that can enhance the quality of interactions between healthcare providers and patients.

Table 2.1   Communication constructs

	Breaking bad news			Conflict situations
A	Anxiety: acknowledge		D	Disagree
K	Knowledge: what do they already know?		A	Agree
I	Information: how much information do they want? Keep it simple, avoid overload		N	Negotiate a compromise
S	Sympathy and emotional management		C	Counsel
S	Support: ask what would help		E	Educate
S	Summarise strategy and key points		R	Refer to third party

Source: Adapted from Mehay et al. (2012).

Table 2.2   Stages of the Calgary–Cambridge framework

Initiating the session	Setting the stage for effective communication by establishing rapport and ensuring a comfortable environment.
Gathering information	Actively listening to the patient's concerns, asking open-ended questions and collecting relevant medical history.
Building the relationship	Developing a trusting and respectful relationship with the patient, acknowledging their emotions and concerns.
Explanations and planning	Clearly conveying information about the diagnosis, treatment options and management plans. Ensuring that the patient understands and is involved in decision-making.
Providing closure	Summarising key points, addressing any remaining concerns and ensuring that the patient is comfortable with the information provided.
Establishing a patient-centred approach	Focusing on the patient's perspective, preferences and understanding and involving them in the decision-making process.
Time and organisation	Managing time effectively and organising the communication process to ensure a comprehensive and patient-centred encounter.
Understanding the patient's perspective	Exploring the patient's beliefs, concerns, expectations and emotional response to their health condition.

Source: Adapted from Kurtz and Silverman (1996).

The key principle underpinning the framework is to combine the pathological description of disease with the patient's subjective experience of illness. Adopting this patient-centred approach has the potential to help establish a trusting relationship between the patient and care provider and to ensure that sufficient information has been gathered for decision-making and lead to a mutually agreed treatment plan.

THE HEALTH HISTORY

A combination of open-ended and specific questions is used during the history-taking process to enable accurate and detailed information to be gathered. The information that is obtained during history taking, along with physical examinations and diagnostic tests, forms the basis for clinical decision-making and the development of a comprehensive and individualised care plan for the patient.

A comprehensive health history commonly consists of several components, such as information about a patient's health, medical background, lifestyle and relevant psychosocial factors. A comprehensive history typically includes the components identified in Table 2.3.

Data concerning sexual and reproductive history for relevant populations may be needed. Obtain information about sexual health, menstrual history, pregnancies and contraception methods.

Table 2.3 Some key components associated with the health history

Health history component	Data gathered	Notes
Gathering data/personal information	• Demographic data • Source of history: patient/carer/medical records	Data provided by the patient is called primary data; data provided by carer/medical records is deemed secondary data
Presenting complaint (PC) Principal symptoms (these may be multiple)	• Major symptoms • Duration • Record each symptom using the patient's own terminology	The chief complaint is the primary reason the patient is seeking medical attention. It is the symptom or set of symptoms that prompted the visit
History of presenting complaint	• Each presenting symptom is explored in detail	
History of presenting illness	• Consider relevant aspects of the review of symptoms	A detailed account of the development of the current symptoms
Present illness	• May include relevant medications, allergies or social influences (alcohol, smoking) that impact the PC • Events are presented in chronological order	Note the onset of the complaint and the progress of the complaint. Factors exacerbating and factors relieving Are there any associated symptoms? Is there a temporal element associated with the complaint (timing)? Are there any episodes of being symptom free?

Health history component	Data gathered	Notes
Past medical history	• List childhood illnesses • List adult illnesses (medical, surgical, obstetric/gynaecological and psychiatric) complete with date of initial diagnosis	A comprehensive review of the patient's medical history. Are there any illnesses associated with morbidity and mortality?
Drug history	• A list of current medications, including prescription and over-the-counter drugs and any supplements. Determine information about medication adherence and any adverse reactions to medications • Allergies and severity. Information about allergies or adverse reactions to medications, foods or environmental factors	Ask the patient to describe the allergic reaction if they have one
Family history	• May be represented in diagram format (a genogram) • Outline age and health or age and cause of death, siblings, parents, grandparents • Consider a genetic cause or contribution to a patient's condition	Note any congenital anomalies and any sibling history A record of the health conditions and diseases that run in the patient's family, helping identify genetic predispositions or familial patterns
Social history	• Occupation and education • Overseas travel • Immunisations • Family of origin • Current household • Personal interests: hobbies • Lifestyle: activities of living, exercise, smoking, alcohol consumption, drug use	Determine what the patient does for work. Do they work with chemicals, dust, animals, paint? Are they exposed to disease exposure?
Review of symptoms	• Review of common symptoms associated with each body system, taking particular note of any red flag symptoms	Make note of and follow up, any red flags identified

Source: Adapted from Diamond-Fox (2023).

Psychosocial history may also be explored. This is in relation to the patient's mental health, stressors, coping mechanisms and support systems.

The role of the healthcare provider requires effective skills in consultation and clinical assessment. This process is intricate, demanding a breadth of knowledge in physiology, pathophysiology and theories of effective communication within the healthcare context. While there are various frameworks available to steer this process, not all may be applicable to the diverse specialist areas in which healthcare providers practice.

APPLICATION OF THE PATIENT HISTORY TO RENAL SYSTEM CONDITIONS

The questions asked should be tailored to the responses the patient makes as well as their individual circumstances. A patient-centred approach, coupled with a detailed assessment, contributes to a more accurate diagnosis. The following is a suggested approach that may be used when obtaining a patient history from people who have or may have suspected renal issues.

INTRODUCTION AND ESTABLISHING RAPPORT

Begin by introducing yourself, explaining your role and establishing a connection with the patient. Take into account the patient's comfort, privacy and any cultural preferences. Use open-ended questions to encourage the patient to share their concerns.

CHIEF COMPLAINT AND PRESENTING SYMPTOMS

Ask the patient about their primary reason for seeking healthcare attention. For those people with suspected renal problems, inquire about symptoms such as:

- Changes in urinary habits (frequency, urgency or nocturia)
- Pain or discomfort in the lower back or sides
- Oedema, especially in the ankles or face
- Changes in urine colour, consistency or odour
- Haematuria
- Fatigue or weakness

HISTORY OF PRESENT ILLNESS

Determine with the patient the onset, duration and progression of symptoms. Ask if there are any factors that trigger or alleviate symptoms. For example:

- When did the symptoms first appear?
- Have they been constant or intermittent?
- Does anything make the symptoms better or worse?

PAIN ASSESSMENT

Use the SOCRATES mnemonic (see Table 2.4), and inquire about the presence, location and characteristics of pain. Renal pain may manifest as dull, aching pain in the lower back or sides.

Ask about factors that worsen or alleviate the pain.

Table 2.4 The SOCRATES in relation to renal pain

Component	Application
Site (S)	Determine the specific site of the pain. In the case of renal pain, patients may point to the lower back, around the area of the kidneys.
Onset (O)	Inquire about when the pain started. For renal issues, it is important to know if the pain began suddenly or developed gradually.
Character (C)	Explore the character or quality of the pain. Renal pain may be described as a dull ache, pressure or sharp pain.
Radiation (R)	Ask whether the pain radiates to other areas. Renal pain might radiate to the lower abdomen or groin.
Association (A)	Inquire about any factors associated with the pain. This could include changes in urinary habits, blood in the urine or other symptoms related to renal function.
Time course (T)	Assess the time course of the pain. Understand if the pain is constant, intermittent or if there are specific triggers or patterns.
Exacerbating/ relieving Factors (E)	Explore factors that exacerbate or alleviate the pain. Renal pain may be influenced by factors such as movement, hydration or certain body positions.
Severity (S)	Have the patient rate the severity of the pain on a numerical scale. This helps gauge the intensity of the renal pain and its impact on the patient.

PAST MEDICAL HISTORY

Inquire about any previous kidney-related issues (urological history) or chronic conditions that could impact renal health, such as:

- Previous kidney stones or urinary tract infections

- Surgery related to the urinary system

- Hypertension or high blood pressure

- Diabetes

- Autoimmune disorders

MEDICATION HISTORY

Ask about the current medications the patient may be taking (prescribed or over the counter), including dosage and frequency. Also, ascertain if the person is using any complementary therapies and, if so, what they are. Pay particular attention to medications that may affect renal function, such as:

- Non-steroidal anti-inflammatory drugs

- Diuretics

- ACE inhibitors or angiotensin II receptor blockers

ALLERGY HISTORY

Ask about any known allergies, especially those related to medications or contrast agents used in medical procedures (for example, iodinated contrast for computed tomography scans, gadolinium-based contrast for MRI). Document the type and severity of allergic reactions.

FAMILY HISTORY

Explore with the patient their family history for conditions that may have a genetic component or impact renal health:

- Kidney disease

- Hypertension

- Diabetes

- Renal diseases or renal replacement therapy

- Inherited conditions affecting the urinary system (for example, polycystic kidney disease)

DIETARY HABITS, INCLUDING SALT AND FLUID INTAKE

Explore the patient's dietary habits, focusing on sodium intake, protein consumption and adherence to any specific dietary restrictions prescribed for renal health. Inquire about the consumption of high-potassium or high-phosphorus foods.

SOCIAL HISTORY

Discuss the patient's lifestyle factors that may influence renal health:

- Dietary habits, including salt and fluid intake

- Exercise routine

- Occupation and potential exposure to nephrotoxic substances

- Smoking history

- Alcohol and substance use

OCCUPATIONAL HISTORY

Inquire about the patient's occupation and any occupational exposures that might contribute to renal issues (nephrotoxic substances). Examples include exposure to chemicals or heavy metals.

REVIEW OF SYSTEMS

Systematically inquire about symptoms related to various organ systems to identify potential systemic conditions affecting the kidneys. For example:

- Cardiovascular: hypertension or cardiovascular disease (ophthalmoscopy/fundoscopy may reveal hypertension)

- Respiratory: shortness of breath or cough

- Gastrointestinal: nausea, vomiting or diarrhoea

- Musculoskeletal: joint pain or swelling

URINARY HABITS

Ask about the patient's urinary habits to gather details on patterns and changes:

- Frequency of urination

- Urgency

- Nocturia (night-time urination)

- Dysuria (painful urination)

- Incontinence

FLUID INTAKE AND OUTPUT

Inquire about the patient's daily fluid intake, including the types and amounts of beverages consumed. Ask about changes in thirst levels and the frequency and volume of urination. Note any signs of dehydration, such as dark urine or infrequent urination. Does the patient look hypovolaemic or fluid-overloaded? Table 2.5 provides information regarding hypovolaemia and fluid overload.

WEIGHT

Sequential measures of a patient's weight are important in the context of renal diseases for several reasons. Monitoring changes in weight can provide valuable information about the patient's fluid status and overall health. This can provide an accurate assessment of fluid loss or gain over the short term.

Table 2.5 Hypovolaemia and fluid overload

Hypovolaemia	Fluid overload
Hypovolaemia, also known as volume depletion, refers to a state of decreased blood volume in the circulatory system. It can result from various conditions that cause the loss of fluids and electrolytes, leading to inadequate perfusion of tissues and organs.	Fluid overload, or hypervolaemia, occurs when there is an excess of fluid in the intravascular and interstitial spaces. This condition can result from several factors.
The patient's eyes may appear sunken and the mucous membranes, dry. Gently, pinch the skin over the anterior chest wall (as opposed to the forearms) to determine if there is reduced skin turgor (elasticity). These features, although relatively insensitive, are most common when there has been significant salt and water loss. This can occur with vomiting or diarrhoea.	The patient with fluid overload may be breathless as a result of pulmonary oedema or pleural effusions and there may be obvious signs of peripheral oedema. Examine for pitting oedema at the base of the spine (sacral oedema, common in bed-bound patients) and in the legs, starting at the ankles, noting the highest level at which oedema can be identified (for example, midcalf, knees or midthigh). In severe cases, oedema can extend into the scrotum or labia. Significant oedema is characteristic of nephrotic syndrome.

Treatment for hypovolaemia and fluid overload involves addressing the underlying cause. In hypovolaemia, intravenous fluids may be administered to restore blood volume, while diuretics and other measures are used to manage fluid overload. Monitoring vital signs, fluid balance and electrolyte levels is crucial in managing these conditions effectively.

Source: Adapted from Dhaun and Kluth (2024).

Trends in weight changes, along with other clinical assessments and laboratory results, can help to guide treatment decisions and interventions for patients with renal diseases. Regular and systematic monitoring of weight is an integral part of managing and providing optimal care for individuals with kidney dysfunction.

FLUID BALANCE CHARTS

The use of fluid balance charts is important in the management of renal diseases as they provide valuable information about a patient's fluid status and aid in optimising care.

The physical examination should be complemented, where possible, by measurement of fluid input (oral and intravenous) and output (urine volumes and other losses, for example, vomitus, diarrhoea). Fluid balance charts provide a systematic and objective way to monitor and manage fluid status, contributing to the prevention of complications and enhancing therapeutic interventions.

PSYCHOSOCIAL HISTORY

Explore psychosocial factors that may impact overall health and well-being:

- Stress levels and coping mechanisms

- Emotional well-being

- Support systems and living arrangements

PATIENT CONCERNS AND EXPECTATIONS

Allow the patient to express their concerns and expectations regarding suspected renal issues. This may include worries about potential diagnoses, treatment options or lifestyle changes. Addressing these concerns enhances patient engagement and collaboration.

It is important to approach each question with sensitivity, acknowledging the potential impact of renal issues on the patient's life. Engage in active listening and be prepared to adapt questioning based on the patient's responses, allowing them to share information at their own pace.

PATIENT EDUCATION NEEDS

Assess the patient's understanding of their renal condition and its management. Identify areas where additional education or clarification may be needed.

Tailor questions based on the patient's responses and individual circumstances. A patient-centred approach, coupled with a detailed assessment, contributes to a more accurate diagnosis and the development of a personalised care plan for individuals with suspected renal issues.

THE PHYSICAL EXAMINATION

When conducting a physical examination of a patient with renal issues, the use of palpation, percussion and inspection is crucial to assess the kidneys, bladder and surrounding structures. Conducting a physical examination of a patient with a renal condition is a crucial aspect of assessing renal health and function. A thorough examination can help identify signs and symptoms related to renal disorders, such as kidney disease, urinary tract infections or other renal issues. Each component of the examination is now discussed.

Begin by introducing yourself to the patient and explaining the purpose of the physical examination. Obtain informed consent, ensuring that the patient understands the nature of the examination and is comfortable proceeding. A chaperone should be offered and provided if requested. Wash your hands and adhere to local policy and procedure regarding infection prevention and control.

Start with a general inspection of the patient. Observe for any signs of distress, fatigue or abnormal postures that may indicate discomfort. Assess the patient's overall appearance, skin colour and hydration status.

Measure and record vital signs, including blood pressure, heart rate, respiratory rate and temperature. Urinalysis can detect the presence of blood, protein, ketones, glucose, leukocytes, nitrites, pH, bile and urobilinogen. The presence of these substances in the urine may indicate different renal disease. They should not, however, be interpreted in isolation, and their presence should prompt further investigation (Gatehouse and Diamond-Fox 2023; see Table 2.6).

Dhauan and Kluth (2024) suggest that urinalysis should be considered an essential part of the renal examination. Urine should be obtained as a midstream specimen so it can be optimally used for subsequent investigations.

Commercially prepared urine dipstick tests use chemical reagents, which change colour when they are immersed in and then removed from urine, to detect abnormalities. Urine test strips contain up to 10 of these chemical pads; however, not all are used in the assessment of renal diseases. It is essential that the manufacturer's instructions are adhered to and the dipsticks are within the expiry date.

Table 2.6   Components of urinalysis

Element	Analysis
Haematuria	Haematuria (renal–glomerulonephritis, pyelonephritis, tumour, infarction, polycystic kidney disease), haemoglobinuria (haemolysis of intravascular origin), myoglobinuria (rhabdomyolysis).
	If present with more than a trace of protein, blood is renal in origin.
	Caution is needed when there is haematuria in the urine. During menstruation, blood is naturally expelled from the uterus, and it should not be confused with blood originating from the urinary system.
Protein	Tests range from + to ++++.
	Only sensitive to albumin.
	Causes of proteinuria may include renal diseases, particularly glomerular in nature. Nonrenal causes can indicate fever, exercise, burns, congestive cardiac failure, post-operative, blood transfusion and acute alcohol abuse.
Ketones	Presence occurs in diabetic ketoacidosis, starvation and low-carbohydrate diets.
Glucose	There may be small amounts present in the urine.
	Presence may indicate diabetes mellitus or the inability of the renal tubules to reabsorb glucose.
Leucocytes	Presence may indicate urinary infection, inflammation, carcinoma.
Nitrates	Presence may indicate gram-negative bacteria.

(Continued)

Table 2.6 (*Continued*)

Element	Analysis
pH	Normally acidic.
	Treatment of specific conditions such as myoglobinuria or urinary calculi may include urinary alkalisation.
	Renal tubular acidosis should be suspected (a condition characterised by a reduced ability of the kidneys to acidify urine properly).
	Infection, e.g. *Proteus mirabilis* (a type of bacteria commonly found in the human gastrointestinal tract and in the environment) may cause alkaline urine.
Bile and urobilinogen	Presence may indicate hepatobiliary disease or haemolysis.

Source: Adapted from Gatehouse and Diamond-Fox (2023); Dhauan and Kluth (2024).

SPECIFIC GRAVITY

The normal range for specific gravity (SG) is 1005–1030.

Specific gravity measures the concentration of solutes (dissolved substances) in urine and provides information about the urine's density compared to the density of water. It is a crucial parameter in urinalysis and can offer insights into the hydration status and the kidneys' ability to concentrate or dilute urine. High SG may indicate dehydration, fever, diabetes mellitus, vomiting or diarrhoea. A low SG may result from over-ingestion of fluid, diabetes insipidus or renal diseases (Swales 2023).

Abnormalities in blood pressure can be indicative of renal issues, especially hypertension. The examination of the patient begins when approaching them, looking for the presence of a catheter bag (if this is present, note urine volume and colour). Determine if there are any alternative vascular access routes in situ and assess whether there are any arteriovenous fistulae present (this determines that the patient has/is receiving renal dialysis).

For examination, the patient should be positioned in a supine position with one pillow under the head and the abdomen should be exposed, but dignity maintained. Ensure comfort, and offer support to the patient if needed.

PALPATION

Kidneys The kidneys are difficult to feel, and a deep bimanual palpation is needed to determine them. Position the patient in a supine or in a slightly tilted lateral position (this involves the patient lying on their side with a subtle inclination, providing an angle between the supine [lying on the back] and lateral [side-lying] positions).

Begin by palpating the right kidney. Place your hands under the patient's right costal margin (the lower border or edge of the ribcage on the right side of the body, where the ribs curve upwards and meet the costal cartilage) and ask them to take a deep breath. Feel for the lower pole of the kidney as it descends during inhalation.

Repeat the process on the left side, palpating the left kidney beneath the left costal margin.

Abdominal Organs   Palpate the entire abdomen systematically, checking for tenderness, masses or abnormalities. Assess for any pulsatile masses, which may indicate an abdominal aortic aneurysm.

Bladder   Palpate the bladder if distension is suspected.

Start above the pubic symphysis (located in the midline of the body at the front of the pelvis, just above the genitals) and palpate gently while asking the patient if they feel fullness or discomfort. If the bladder is full, the palpation will reveal an oval shape which is firm, smooth and regular in nature.

PERCUSSION

Kidneys   Percuss the kidneys to identify any areas of dullness.

Place one hand over the costovertebral angle (located on the back, on both sides of the spine, where the lower ribs meet the vertebral column) and percuss with the other hand. A dull sound may indicate kidney pathology.

Figure 2.1 provides a diagrammatic representation of the kidney palpation.

Bladder   Percuss the bladder to assess for distension.

Begin percussing at the midline just above the pubic symphysis. A dull sound indicates a full bladder.

INSPECTION

Kidneys   Inspect the lower back for any asymmetry, masses or abnormalities.

Look for any visible pulsations or vascular abnormalities.

Abdomen   Observe the abdomen for any distension, asymmetry or visible masses.

Note any visible pulsations or abnormalities in the abdominal wall.

Skin   Inspect the skin for signs of renal diseases, such as pallor, jaundice or ecchymosis (bruising, discolouration of the skin).

Look for evidence of pruritus or uremic frost (the formation of a powdery, crystalline material on the skin's surface), particularly in advanced renal diseases.

Oedema   Inspect for peripheral oedema, especially in the lower extremities, which may be associated with fluid retention in renal disorders.

ADDITIONAL FACTORS

- Always communicate with the patient throughout the examination, explaining each step and ensuring their comfort.

- Assess for any signs of pain or discomfort during palpation and percussion, as this can provide valuable information.

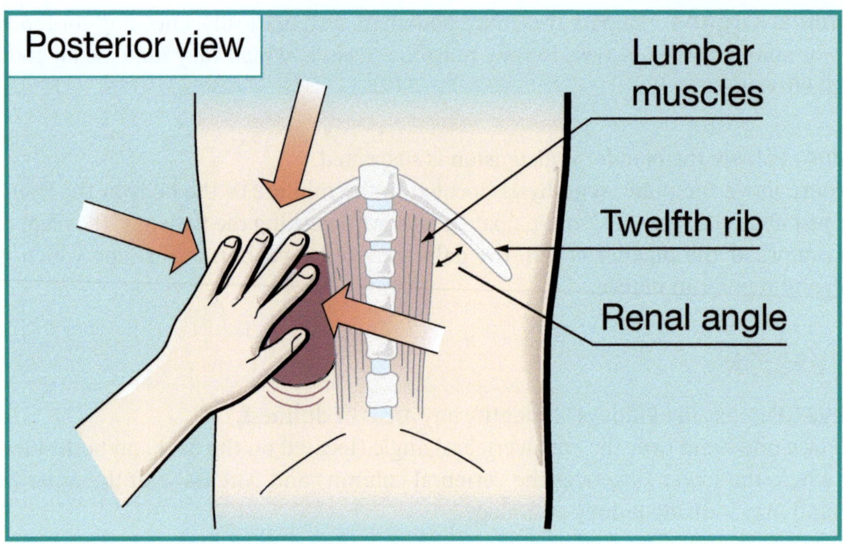

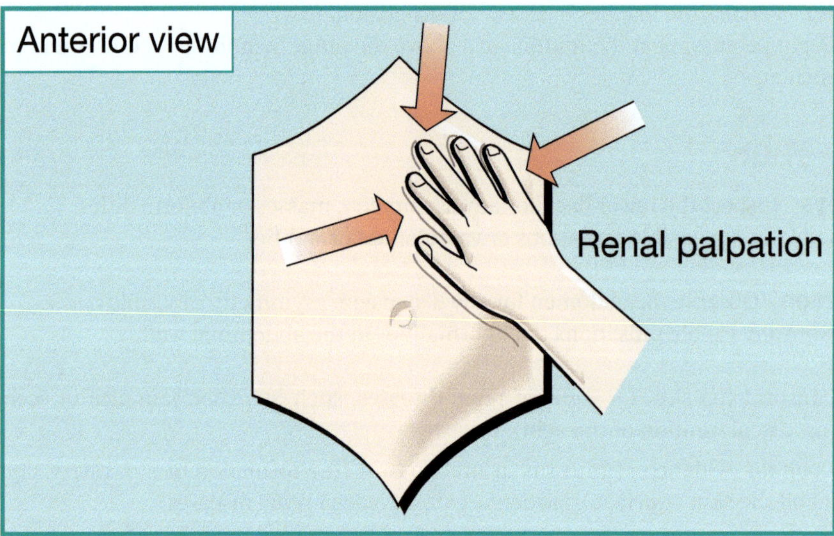

FIGURE 2.1 Kidney palpation

- Consider the patient's overall hydration status, as dehydration can affect the palpation findings.

- Document any relevant findings thoroughly in the patient's medical record and adhere to local policy and procedure.

- It is important to note that the physical examination is part of a broader assessment and abnormal findings may warrant further investigation through laboratory tests, imaging studies or consultation with specialists. Collaborate with other healthcare professionals, such as nephrologists or urologists, as needed for a comprehensive evaluation of renal health.

Figure 2.2 provides a diagrammatic representation of the history and examination related to renal diseases.

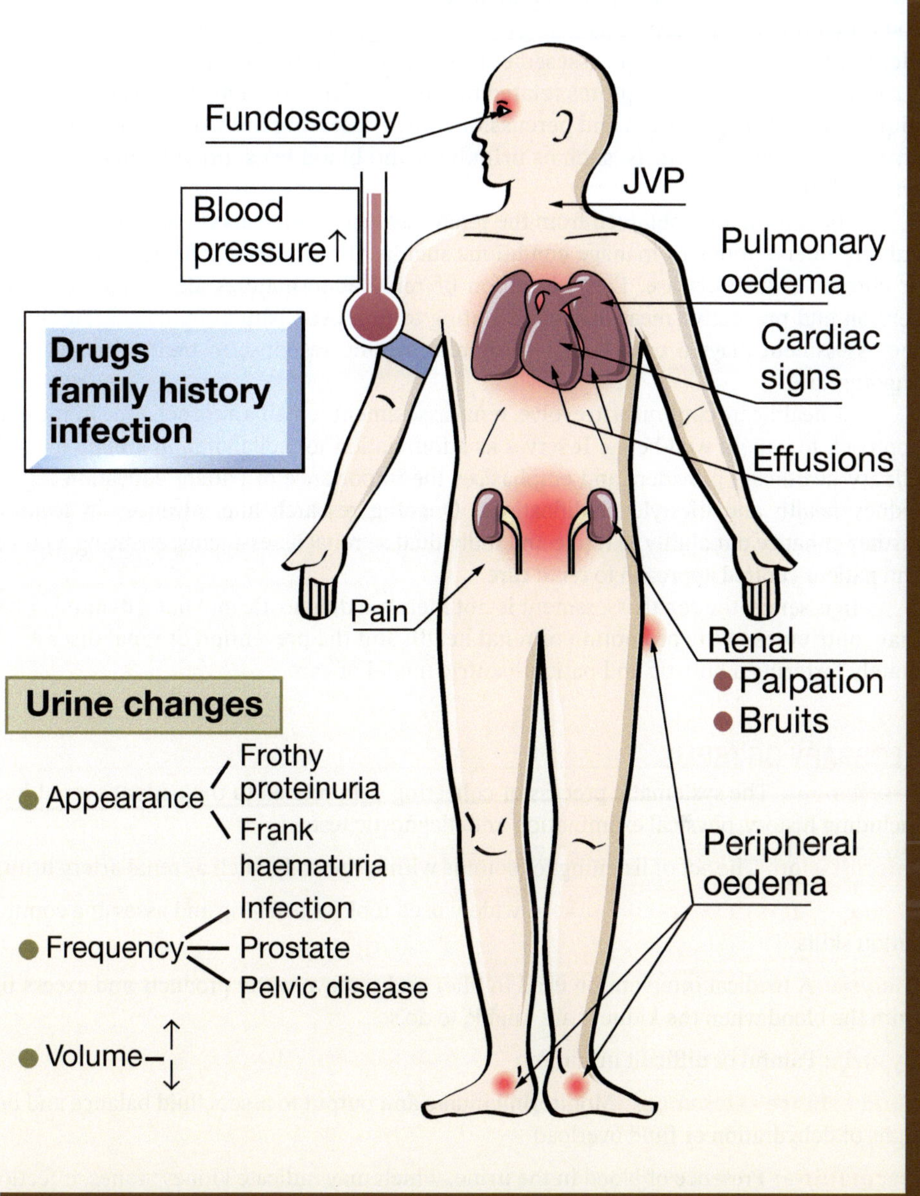

FIGURE 2.2   History and examination related to renal disease

CONCLUSION

A thorough assessment of the renal system is essential for the provision of patient-centred care. The integration of subjective and objective data provides a comprehensive understanding of the patient's renal health, facilitating early detection of dysfunction and implementation of timely interventions. Regular monitoring, collaboration with the healthcare team and patient education contribute to the holistic care of individuals with renal concerns.

The renal assessment is a critical component of comprehensive patient care, offering valuable insights into kidney function, fluid balance and overall health. Through a systematic approach incorporating history taking, physical examination and diagnostic tests, healthcare professionals can gain a thorough understanding of renal health and identify potential issues. The assessment involves evaluating factors such as urinary patterns, fluid intake and symptoms related to renal dysfunction. Physical examination techniques, including palpation and percussion, aid in assessing the kidneys and surrounding structures. Diagnostic tests, such as urinalysis and blood tests, provide quantitative data on renal function.

The information obtained from the renal assessment informs clinical decision-making, guiding interventions to manage conditions such as urinary tract infections, kidney stones or chronic kidney disease. Early detection of renal abnormalities allows for timely intervention and preventive measures, contributing to improved patient outcomes. Additionally, the assessment plays a crucial role in monitoring the response to treatment and guiding ongoing care.

As healthcare continues to evolve, renal assessment remains a cornerstone in the holistic approach to patient well-being. It serves as a foundation for collaboration among multidisciplinary healthcare providers and emphasises the importance of patient education regarding kidney health and lifestyle modifications. Ongoing research and advances in technology further enhance our ability to refine and individualise renal assessments, ensuring a proactive and patient-centred approach to renal care.

In essence, the renal assessment is not merely a diagnostic tool but a dynamic process that contributes to the promotion of renal health and the prevention of renal diseases, ultimately fostering a holistic and patient-centric model of care.

GLOSSARY OF TERMS

Assessment: The systematic process of collecting data related to the patient's renal health, including history, physical examination and diagnostic tests.

Auscultation: The act of listening for sounds within the body, such as renal artery bruits.

Calgary–Cambridge framework: A widely used tool for teaching and assessing communication skills.

Dialysis: A medical intervention used to filter and remove waste products and excess fluids from the blood when the kidneys are unable to do so.

Dysuria: Painful or difficult urination.

Fluid balance assessment: Monitoring intake and output to assess fluid balance and detect signs of dehydration or fluid overload.

Haematuria: Presence of blood in the urine, which may indicate kidney stones, infection or other renal issues.

History taking: Collecting a comprehensive patient history, including information about urinary habits, medications and risk factors.

Inspection: Visually examining the abdomen and lower back for signs of abnormalities or kidney enlargement.

Mnemonic: Memory aid or learning technique that helps people remember information more easily.

Nephrotoxic: Substances or medications that have the potential to cause kidney damage.

Nocturia: Passing urine at night.

Observation: Monitoring for signs and symptoms of renal dysfunction, such as changes in urine colour, frequency or volume.

Palpation: Assessing the kidneys for tenderness, masses or other abnormalities through gentle palpation of the abdomen.

Uraemia: Build-up of waste products in the blood due to impaired kidney function, leading to symptoms like nausea, fatigue and confusion.

Urinalysis interpretation: Analysing the results of a urinalysis, including aspects such as colour, specific gravity and the presence of protein or blood.

Vital sign monitoring: Regularly assessing vital signs, including heart rate and blood pressure, to detect changes in haemodynamic status.

MULTIPLE CHOICE QUESTIONS

1. What is the primary function of the kidneys?
    a) Digestion
    b) Filtration and excretion of waste products
    c) Respiratory gas exchange
    d) Immune system regulation

2. What is the term for the presence of glucose in the urine?
    a) Albuminuria
    b) Glycosuria
    c) Haematuria
    d) Pyuria

3. Which condition is characterised by the presence of blood in the urine?
    a) Haematuria
    b) Glycosuria
    c) Oliguria
    d) Anuria

4. Why is the measurement of urine specific gravity important in the assessment of kidney function?
    a) To assess for signs of dehydration
    b) To monitor acid–base balance
    c) To detect signs of urinary tract infection
    d) To evaluate kidney perfusion

5. What is the primary goal of effective communication during a clinical assessment?
    a) Sharing personal experiences
    b) Building rapport with the patient
    c) Transmitting information accurately
    d) Using complex medical terminology

6. Why is it important to use open-ended questions during a clinical assessment?
    a) To save time
    b) To gather specific information
    c) To encourage detailed responses
    d) To control the conversation

7. What is the purpose of providing a summary or clarification during communication with a patient?
   a) To confuse the patient
   b) To repeat information unnecessarily
   c) To show off medical knowledge
   d) To ensure mutual understanding

8. Which assessment technique is appropriate for detecting kidney tenderness or pain in a patient with renal disease?
   a) Percussion
   b) Palpation
   c) Auscultation
   d) Inspection

9. In a patient with kidney disease, what does the term 'uraemia' indicate?
   a) Increased levels of nitrogen-containing compounds in the blood
   b) Normal kidney function
   c) Decreased levels of creatinine in the blood
   d) Presence of glucose in the urine

10. Why is the measurement of urine output important in the assessment of kidney function?
    a) To evaluate cardiac function
    b) To monitor acid–base balance
    c) To detect signs of urinary tract infection
    d) To assess kidney perfusion and function

REFERENCES

Dhauan, N. and Kluth, D. (2024). The renal system (Chapter 12). In: *Macleod's Clinical Examination*, 15e (eds. A.R. Dover, J.A. Innes, and K. Fairhurst). London: Elsevier.

Diamond-Fox, S. (2023). Undertaking consultations and clinical assessments at advanced level. *British Journal of Nursing*. 30 (4): 238–243.

Dover, A.R., Fairhurst, K., and Innes, J.A. (2024). General aspects of examination (Chapter 3). In: *Macleod's Clinical Examination*, 15e (eds. A.R. Dover, J.A. Innes, and K. Fairhurst). London: Elsevier.

Fairhurst, K., Innes, J.A., and Dover, A.R. (2024). Managing clinical encounters with patients (Chapter 1). In: *Macleod's Clinical Examination*, 15e (eds. A.R. Dover, J.A. Innes, and K. Fairhurst). London: Elsevier.

Fawcett, T. and Rhynas, S. (2012). Taking a patient history: the role of the nurse. *Nursing Standard*. 26 (24): 41–46.

Gatehouse, A. and Diamond-Fox, S. (2023). Renal critical care (Chapter 24). In: *Fundamentals of Critical Care*. (eds. I. Peate and B. Hill). Oxford: Wiley.

Kurtz, S.M. and Silverman, J.D. (1996). The Calgary–Cambridge referenced observation guides: an aid to defining the curriculum and organising the teaching in communication training programmes. *Medical Education*. 30 (2): 83–89.

Mehay, R., Beaumont, R., Draper, J. et al. (2012). Revisiting models of the consultation (online-only content. In: *The Essential Handbook for GP Training and Education*. (ed. R. Mehay). London: Radcliffe Publishing. https://tinyurl.com/yaxqupm3.

Swales, J. (2023). The person with a urinary disorder (Chapter 29). In: *Nursing Practice*, 3e (eds. I. Peate and A. Mitchell). Oxford: Wiley.

Pyelonephritis CHAPTER 3

Pyelonephritis is from the Greek word 'pyelo' (pelvis), 'nephros' (kidney) and '-itis' (inflammation). Acute pyelonephritis is an infection within the renal pelvis, usually accompanied by infection within the renal parenchyma (renal tissue); it can occur in one or both kidneys. The source of the infection is often ascending infection from the bladder, but haematogenous spread can also occur (see Box 3.1, regarding haematogenous spread). The usual organisms are the same as for lower urinary tract infection, for example, *Escherichia coli*, *Klebsiella* spp., *Proteus* spp., *Enterococcus* spp. (see Table 3.1). Unusual organisms are occasionally seen – such as mycobacteria, yeasts and fungi and opportunistic pathogens such as *Corynebacterium urealyticum*. Repeated attacks of acute pyelonephritis can lead to chronic pyelonephritis, which involves the destruction and scarring of renal tissue due to repeated inflammation.

BOX 3.1	HAEMATOGENOUS SPREAD

Haematogenous spread refers to the dissemination or transmission of microorganisms (such as bacteria or viruses) through the bloodstream. This method of transmission is particularly relevant in the context of infections.

When an infection occurs, the pathogen can enter the bloodstream and can then be carried to different parts of the body. This is in contrast to localised infections, where the infection is limited to a specific area or organ. Haematogenous spread allows microorganisms to reach distant sites in the body, leading to the potential development of infections in areas that are not directly connected to the initial site of infection.

In the context of pyelonephritis, haematogenous spread may not be the primary mode of infection. Pyelonephritis typically arises from ascending infection, where bacteria from the lower urinary tract ascend through the ureters to infect the kidneys. However, in some cases, particularly in severe or complicated infections, bacteria can enter the bloodstream and be carried to the kidneys, leading to haematogenous spread of the infection.

Haematogenous transmission is a significant factor in the dissemination of infections and can contribute to the development of systemic infections or the involvement of distant organs. Understanding how infections are transmitted through the bloodstream is crucial for diagnosing and treating various infectious diseases.

PATHOPHYSIOLOGICAL CHANGES ASSOCIATED WITH PYELONEPHRITIS

A urinary tract infection is a broader term encompassing infections in any part of the urinary system; pyelonephritis, however, specifically refers to a kidney infection. Pyelonephritis is generally considered a more severe form of urinary tract infection, and it may require more aggressive treatment, especially if complications arise. Both conditions, if they are left untreated, can lead to serious health complications. See Table 3.2 for a comparison of urinary tract infections and pyelonephritis.

Table 3.1 Microorganisms commonly associated with pyelonephritis

Organism	Discussion
Escherichia coli	*E. coli* is a species of bacteria commonly found in the lower intestines. Most strains of *E. coli* are harmless and a part of the normal flora of the human gut, where they contribute to the maintenance of a healthy intestinal environment. However, certain strains of *E. coli* can cause illness.
	E. coli is a frequent cause of pyelonephritis and is responsible for a significant proportion of uncomplicated community-acquired and healthcare-associated urinary tract infections.
Klebsiella spp.	*Klebsiella* spp. is a group of bacteria that belongs to the family Enterobacteriaceae. They are normally found in the environment and in the intestinal tract, where they do not usually cause disease. However, if they are transmitted to other parts of the body, for example, the urinary tract, they can cause serious infections.
Proteus spp.	*Proteus* is a genus of gram-negative bacteria that is known for their motility and diverse characteristics. The two primary species within the *Proteus* genus that are medically relevant are *Proteus mirabilis* and *Proteus vulgaris*. These bacteria are part of the normal intestinal flora but can also be associated with various infections, particularly *P. mirabilis*.
Enterococcus spp.	*Enterococci* are gram-positive microorganisms that can thrive in environments with or without oxygen and they are part of the normal flora of the human gastrointestinal tract. While they are generally commensal bacteria, meaning they coexist harmlessly with their human host, some species can cause infections, especially in individuals with compromised immune systems or underlying health conditions.
Pseudomonas spp.	*Pseudomonas* is a genus of bacteria that includes various species known for their ability to thrive in diverse environments. *Pseudomonas aeruginosa* is one of the most clinically significant species within this genus, as it is a common opportunistic pathogen that can cause infections in humans, particularly in individuals with compromised immune systems or underlying health conditions.

Source: Adapted from UK Health Security Agency (2017); British Medical Journal (2023).

Figure 3.1 offers a diagrammatic representation of urinary tract infection.

The pathophysiology of pyelonephritis is complex and comprises several mechanisms. The pathogenesis of pyelonephritis involves the ascent of bacteria from the lower urinary tract to the kidneys, resulting in infection and inflammation of the renal parenchyma and pelvis. Table 3.2 offers an overview of the pathogenesis of pyelonephritis.

As bacteria enter the urinary tract, this initiates a cascade of events that leads to infection and inflammation of the kidneys. Microorganisms (see Table 3.2) gain access through the urethra, often during activities such as sexual intercourse. Other factors, such as urinary tract abnormalities, pregnancy or catheter use, heighten vulnerability to infection.

Once in the urinary tract, the bacteria use specific adhesion factors that help them stick or attach to surfaces. Type 1 pili in *E. coli* are specific structures located on the surface of *E. coli* bacteria, acting like a tiny hook that helps the bacteria attach to things, allowing them to cling to uroepithelial cells (cells lining the urinary tract). This ability to adhere to the cells is part of how bacteria can cause urinary tract infections. Adhesion is important for colonisation and subsequent infection.

Table 3.2 A comparison of pyelonephritis and urinary tract infection

	Pyelonephritis		Urinary tract infection
Defined as	Specifically refers to an infection of the kidneys.	Defined as	A bacterial infection that affects any part of the urinary system, including the bladder, urethra, ureters and in some cases, the kidneys.
Location	An infection that involves the renal parenchyma (tissue) and pelvis of one or both kidneys.	Location	It can occur in the lower urinary tract (bladder and urethra) or less commonly, in the upper urinary tract (kidneys).
Symptoms	It often includes fever, chills, back or flank pain, nausea, vomiting and urinary symptoms similar to those of a lower urinary tract infection. The symptoms are generally more severe compared to a typical lower urinary tract infection.	Symptoms	It includes frequent micturition and dysuria (painful urination), a strong urge to urinate and cloudy or malodorous urine. There may be discomfort or pressure in the lower abdomen, fever, back or flank pain and nausea.
Causes	The most common cause is the ascent of bacteria, usually from the lower urinary tract. *E. coli* is the predominant pathogen, but other bacteria may also be involved.	Causes	Most are caused by bacteria; *E. coli* is the most common pathogen. Other bacteria, such as *Proteus*, *Klebsiella* and *Enterococcus*, can also be responsible.
Risk factors	A history of recurrent urinary tract infections, urinary tract abnormalities, kidney stones and conditions that compromise the immune system.	Risk factors	Female gender (due to a shorter urethra), sexual activity, change in sexual partner, urinary tract abnormalities, catheter use and conditions that compromise the immune system.

Source: Adapted from Hudson and Mortimore (2020).

Bacteria, especially *E. coli*, have the capability of entering the cells that line the urinary tract, where they create reservoirs (hiding spots or storage spaces) and this contributes to recurrent infections (the infections happen again and again). The ability of the bacteria to hide inside the cells makes it more challenging for the immune system or antibiotics to completely eliminate them, leading to recurrent infections. Bacterial virulence factors, these include toxins and adhesins (a molecule or a structure on a cell), enhance their ability to cause infection and evade host defences.

The bacteria continue to ascend from the bladder to the kidneys. Vesicoureteral reflux and urinary tract obstructions facilitate this ascent (see Figure 3.1), creating conditions that promote the upward movement of bacteria in the urinary system, increasing the risk of kidney infections. Upon reaching the kidneys, the infection triggers an inflammatory response. Pro-inflammatory mediators are released, and they send out signals, prompting the recruitment of immune cells, predominantly neutrophils, to the site of infection. These immune cells play a vital role in combating invading pathogens and facilitating the resolution of the infection.

Inflammatory processes and the presence of bacteria result in tissue damage within the kidneys. Abscesses may form, representing localised collections of pus (pyonephrosis). Complications can arise, including sepsis and in chronic cases, scarring with impaired kidney function.

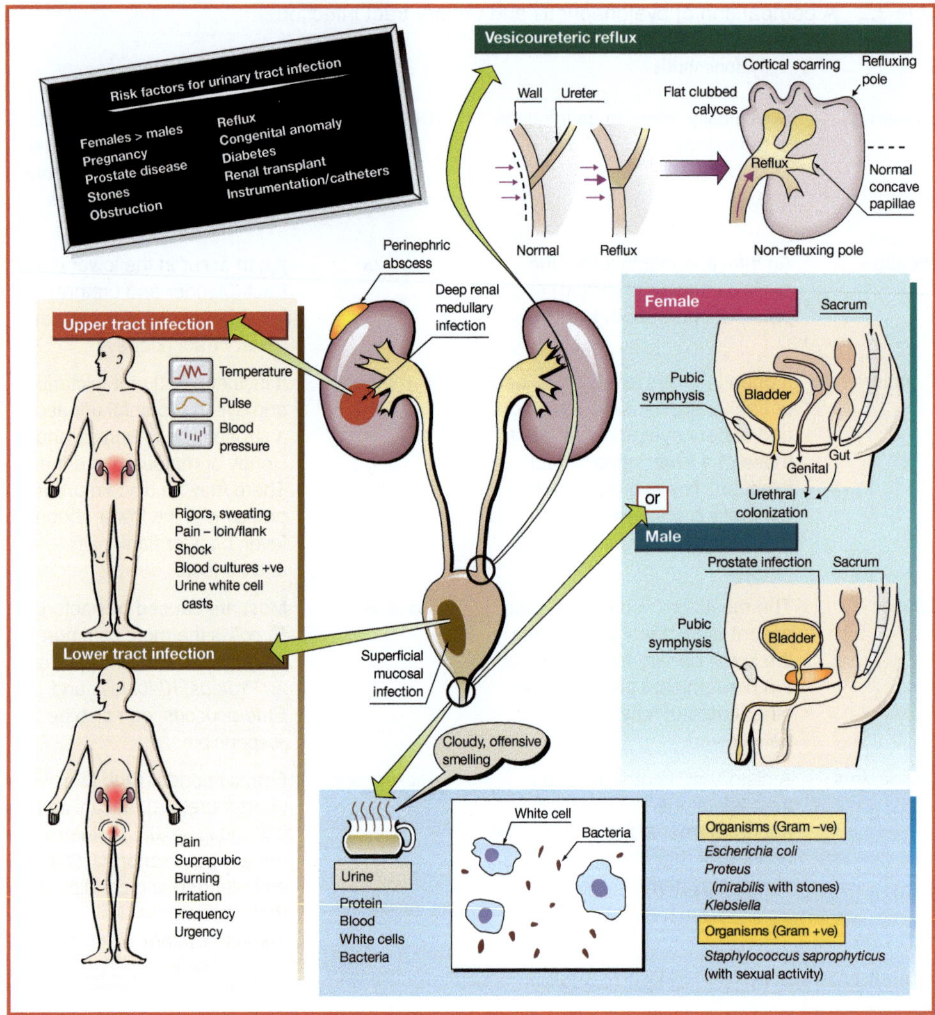

FIGURE 3.1   Urinary tract infection

This understanding of pyelonephritis pathogenesis informs comprehensive approaches to diagnosis, treatment and prevention. Table 3.3 provides a summary of the pathogenesis associated with pyelonephritis.

Urinary tract obstruction caused by a kidney stone, for example, may also lead to acute pyelonephritis. An outflow obstruction of urine can result in incomplete emptying and urinary stasis, which in turn causes bacteria to multiply without being flushed out.

EPIDEMIOLOGY

Young sexually active women are those who are most often affected by acute pyelonephritis. Groups with extremes of age, for example, the elderly and infants, are also at risk due to changes in their anatomy and changes in hormones. Pregnant women can also be at risk; around 20–30% will develop acute pyelonephritis, usually during the second and early third trimester. Acute pyelonephritis has no racial predisposition (Matuszkiewicz-Rowińska, Małyszko, and Wieliczko 2015).

Table 3.3 Pathogenesis associated with pyelonephritis: a summary

Pathophysiology	Discussion
Bacterial entry into the urinary tract	The majority of urinary tract infections and subsequently pyelonephritis are caused by bacteria entering the urethra, and they ascend through the urinary tract.
	Factors contributing to bacterial entry include sexual activity, urinary catheterisation, urinary stasis (reduced urine flow) and the presence of anatomical abnormalities that may facilitate bacterial ascent.
Adherence to uroepithelial cells	Bacteria possess specialised surface structures, such as fimbriae or pili, which resemble hair-like or thread-like appendages. These features act like hooks, helping the bacteria adhere to the cells that line the urinary tract. As these bacteria encounter the cells (called uroepithelial cells) lining the urinary tract, they can latch on and adhere to them. This adherence is important as the bacteria can establish an infection in the urinary tract.
Invasion of uroepithelial cells	Some bacteria can invade uroepithelial cells. Intracellular invasion allows bacteria to evade host immune responses and establish a persistent infection.
	Intracellular bacteria can form quiescent reservoirs, contributing to recurrent infections.
Initiation of the inflammatory response	Intracellular invasion allows bacteria to avoid host immune responses and establish a persistent infection. Intracellular bacteria can form reservoirs that contribute to recurrent infections.
	Bacterial invasion triggers the release of inflammatory mediators, which then leads to the activation of the host immune response.
	Inflammatory cells, such as neutrophils, are recruited to the site of infection.
Renal involvement and formation of abscesses	Bacteria can ascend from the bladder to the kidneys, leading to pyelonephritis.
	Infection and inflammation extend into the renal parenchyma, causing damage to the kidney tissue and leading to the formation of abscesses.

Acute pyelonephritis can occur at any age. Around 1% of boys and 3% of girls will have had acute pyelonephritis by the age of 7 years. Incidence is highest in women aged 15–29 years, followed by infants and older people (Colgan, Williams, and Johnson 2011). In men, this is relatively uncommon. Over the age of 65 years, the incidence in men rises to match that of women.

Hudson and Mortimore (2020) note that annually, one in every 830 people in England is estimated to develop pyelonephritis. However, Bethel (2012) suggests that the exact epidemiology and economic cost remain unknown because of the overlap of treatment in primary and secondary care.

RISK FACTORS

Pyelonephritis can affect anyone. However, there are some people who are more at risk than others, for example, pregnant women, people with diabetes, people with kidney stones and those with a weakened immune system.

Women face a higher risk due to the anatomical distinction of having a shorter urethra; this makes it easier for bacteria to ascend from the lower urinary tract to the kidneys.

Structural abnormalities in the urinary tract, such as kidney stones or obstructions, can create conditions conducive to infection. Individuals with a history of recurrent or untreated lower urinary tract infections are at an increased risk of developing pyelonephritis.

Age is also a contributing factor, with both infants and the elderly being more susceptible (the extremes of age). Infants may be at risk due to congenital abnormalities, while the elderly often experience weakened immune systems and may have comorbidities. Sexual activity, particularly in women, can introduce bacteria into the urethra, increasing the likelihood of urinary tract infections that may progress to pyelonephritis. Pregnancy induces physiological changes in the urinary tract, making pregnant women more prone to these infections.

Medical conditions such as diabetes, which can compromise the immune system, can heighten susceptibility to pyelonephritis. Similarly, individuals with suppressed immune systems, either due to conditions such as HIV/AIDS or immunosuppressive medications, can face an increased risk. The use of urinary catheters can introduce bacteria into the urinary tract, contributing to infection.

Anatomical issues, such as vesicoureteral reflux, where urine flows backwards from the bladder to the kidneys, create a pathway for bacteria to reach the kidneys. In men, an enlarged prostate gland may obstruct urine flow, increasing the risk of urinary tract infections and subsequent pyelonephritis. Neurological disorders affecting nerve function, for example, spinal cord injuries, can disrupt normal bladder function, raising the likelihood of urinary tract infections.

Timely diagnosis and appropriate treatment of urinary tract infections are crucial in preventing the progression of pyelonephritis.

Understanding the risk factors associated with pyelonephritis plays a pivotal role in preventive measures. Insight and awareness of these risk factors can help healthcare providers implement strategies to reduce the likelihood of developing pyelonephritis.

Awareness of risk factors contributes to early detection. By recognising symptoms or signs early on, healthcare providers can promptly diagnose pyelonephritis and initiate timely treatment. This proactive approach helps prevent the infection from advancing and causing more severe complications.

Recognising the various risk factors can help focus on those populations that may be at a higher risk, leading to the early identification of infections, especially in susceptible groups. Tailoring treatment is another key aspect influenced by an understanding of risk factors. Different risk factors may require specific interventions or adjustments to the treatment plan. This individualised approach contributes to more effective and personalised care. See Table 3.4 for an overview of risk factors.

CLINICAL PRESENTATION

There are no clinical features or routine investigations that conclusively distinguish acute pyelonephritis from lower urinary tract infection. Clinical presentation can vary and not all individuals with pyelonephritis will experience all of these symptoms. Additionally, in some cases, individuals, especially older adults or those with weakened immune systems, may present with subtle or atypical symptoms.

One of the cardinal signs is fever. Often this can be accompanied by chills and an indication of the body's robust immune response to the infection. A distinguishing feature of pyelonephritis is the presence of flank pain, a persistent and dull ache localised on one or both sides

Table 3.4 Risk factors associated with pyelonephritis

Gender	Women are at a higher risk than men
Urinary tract abnormalities	Structural abnormalities in the urinary tract, such as kidney stones or urinary tract obstructions.
Urinary tract infections	Recurrent or untreated lower urinary tract infections.
Age	Infants and elderly individuals are at a higher risk.
Sexual activity	Sexual activity, particularly in women.
Pregnancy	Pregnant women are at an increased risk due to changes in the urinary tract and hormonal influences.
Diabetes	Individuals with diabetes are more susceptible to pyelonephritis.
	Poorly controlled blood glucose weakens the immune system, promoting bacterial growth. Nerve damage in diabetes may lead to incomplete bladder emptying, which creates a breeding ground for bacteria.
Immunosuppression	Conditions or medications suppressing the immune system.
Catheter use	Urinary catheterisation for an extended period.
Vesicoureteral reflux	Backward flow of urine from the bladder to the kidneys provides a pathway for bacteria to reach the kidneys.
Prostatic hypertrophy	Enlarged prostate.
Neurological disorders	Conditions that affect nerve function.

Source: Adapted from Hagler et al. (2023).

of the lower back, beneath the ribs. This pain results from the inflammation of the kidneys and this is a key indicator of the infection's presence.

Abdominal pain is another common symptom, which reflects the broader impact of the infection on the renal system. Changes in urinary habits are typical, including heightened urgency and frequency of urination. Discomfort or a burning sensation during urination is also prevalent (dysuria). Particularly, the urine may exhibit cloudiness and an unpleasant odour due to the presence of pus and bacteria. Haematuria may be revealed as a pink or reddish hue, highlighting the inflammatory response that is affecting the urinary tract.

In more severe cases, individuals may experience systemic symptoms such as nausea and vomiting, indicating the body's widespread reaction to the infection (European Association of Urology 2023). Fatigue and a general sense of weakness are also common. These stem from the inflammatory processes as well as the physiological toll of combating the infection. In elderly individuals, pyelonephritis may present with subtle or atypical symptoms, which can include confusion or altered mental status/cognition (European Association of Urology 2023).

It is critical to recognise the variability in the clinical presentation of pyelonephritis, with not all individuals experiencing every symptom. The manifestation of subtle or atypical symptoms is particularly relevant for older adults or those with a compromised immune system. Early diagnosis and the administration of appropriate antibiotics are paramount to prevent complications and enable a successful recovery from pyelonephritis.

CLINICAL INVESTIGATIONS AND DIAGNOSIS

Acute pyelonephritis is diagnosed by taking a detailed medical history, performing a physical examination, and undertaking a combination of clinical investigations and assessments to confirm the presence of a kidney infection.

HISTORY TAKING

A comprehensive history-taking activity is key for making a diagnosis of pyelonephritis. Healthcare providers engage in a systematic interview with the patient to gather relevant information.

Effective communication is a key aspect underpinning the history-taking process. Establishing clear and open communication between the healthcare provider and the patient is crucial for obtaining accurate information and for building a trusting relationship. Key aspects of communication in the history-taking process are described in Table 3.5.

Effective communication will not only aid in accurate information gathering but can also play a crucial role in patient satisfaction and adherence to treatment plans. It is a fundamental skill that healthcare providers must continuously refine to enhance the quality of care.

In the context of history taking for suspected pyelonephritis, there are specific key aspects required to gather relevant information for diagnosis and treatment. Consider the patient's symptomatology, specifically asking the patient about the presence of fever, chills and flank pain, which are classic indicators of kidney infection. Detailed questions are directed at urinary symptoms, such as:

- Increased urgency

- Frequency of micturition

- Painful urination

- Any changes in urine appearance

- The presence of blood or pus

A crucial aspect involves exploring the patient's recent history of urinary tract infections and any antibiotic use. Individuals with recurrent urinary tract infections may be at a higher risk of developing pyelonephritis. Ask the patient about potential risk factors, such as diabetes, immunosuppression, recent catheter use or urinary tract abnormalities that might compromise the immune system.

For female patients of reproductive age, determining pregnancy status is vital, considering its impact on susceptibility to urinary tract infections and treatment decisions. Additionally, a travel history may be explored, especially if the patient has been to regions where specific infectious agents are more prevalent.

The healthcare provider investigates complications and associated symptoms, such as nausea, vomiting or confusion, to assess the severity of the infection and identify potential complications such as sepsis or kidney damage. Patients with a history of pyelonephritis may be asked about previous occurrences to tailor treatment plans and preventive measures accordingly. Inquiring about any previous diagnostic tests or treatments for the current symptoms provides additional context for the healthcare provider.

Gathering information on these specific aspects aids healthcare providers in narrowing down the differential diagnosis, formulating an appropriate plan for further investigations and devising tailored treatment strategies for suspected pyelonephritis.

Table 3.5   Key aspects of communication in the history-taking process

Aspect	Discussion
Establishing rapport	Building a rapport with the patient creates a comfortable and trusting environment. This can encourage patients to share their symptoms and medical history openly.
Active listening	Active listening involves giving full attention to the patient, acknowledging their concerns and responding appropriately. This helps in understanding the patient's perspective and gathering detailed information.
Empathy	Demonstrating empathy and understanding towards the patient's experiences fosters a positive connection. It encourages patients to share sensitive information and enhances the overall quality of communication.
Clarity and simplicity	Using clear and simple language helps ensure that patients understand the questions being asked. Avoiding medical jargon and providing explanations when necessary contributes to effective communication.
Open-ended questions	Encourage patients to provide detailed responses by asking open-ended questions. This allows for a more thorough exploration of symptoms and relevant information.
Non-verbal communication	Non-verbal cues, such as body language and facial expressions, play a significant role in communication. Healthcare providers should be attentive to both verbal and non-verbal signals to gain a holistic understanding.
Cultural sensitivity	Recognising and respecting cultural differences is vital. Cultural sensitivity ensures that questions are framed in a way that is culturally appropriate, facilitating better communication and understanding.
Privacy and confidentiality	Emphasising the importance of privacy and confidentiality helps create a safe space for patients to share personal and sensitive information without fear of judgement or a breach of privacy.
Encouraging patient participation	Encouraging patients to actively participate in the conversation has the potential to empower them to share their perspectives on symptoms, contributing to a more collaborative diagnostic process.
Clarifying doubts	Patients may have concerns or questions about their symptoms or the diagnostic process. Allowing time to address these concerns and provide clarifications fosters a sense of involvement and trust.
Summarising and confirming information	Periodically summarising the information obtained and confirming it with the patient ensures that there is a shared understanding. This may help in avoiding misunderstandings and contribute to accurate history taking.
Adaptability	Recognising that each patient is unique and may require different communication approaches is essential. Being adaptable in communication style ensures that information is effectively exchanged with diverse patient populations.

Source: Adapted from Hudson and Mortimore (2020).

THE PHYSICAL EXAMINATION

Physical examination findings specific to pyelonephritis help corroborate the clinical suspicion. The physical examination is a critical component of the healthcare assessment process; it provides valuable information about a patient's overall health and aids in the diagnosis of specific conditions.

Before the examination, review the patient's medical history, including past illnesses, medications, allergies and relevant details. Create a comfortable environment and ensure the examination room is well-lit, warm and the patient is comfortable. Explain the process to the patient, addressing any concerns they may have. The presence of a chaperone should be offered. Local policy and procedure concerning infection prevention and control much be adhere to and informed consent gained.

Adapt your approach to meet patient needs, considering patient comfort at all times. There may be a need to modify the examination based on the patient's comfort level, ensuring they feel at ease throughout the process. Adapt the examination for special populations, for example, older people, those with limited mobility or differently abled patients, taking into account their unique needs and considerations.

Prepare the necessary equipment, such as a stethoscope, blood pressure cuff, thermometer and other specific instruments relevant to the examination.

When approaching the physical examination, this begins with a general observation of the patient. Note the patient's posture, skin colour, signs of distress or any obvious abnormalities. Measure vital signs, including blood pressure, heart rate, respiratory rate and body temperature. These provide baseline information about the patient's overall health.

Document and communicate findings in accordance with local policies and procedures, using a systematic approach. Include any abnormalities, measurements, and specific details observed during the examination. Share relevant findings with the patient, explain any abnormalities and address their questions or concerns.

It may also be important to share examination findings with other members of the healthcare team. This approach facilitates the production of a comprehensive care plan with the patient central to all that is done.

A systematic and thorough physical examination, coupled with effective communication, collaboration and a detailed patient history, contributes to a comprehensive understanding of the patient's health and aids in the formulation of an accurate diagnosis and treatment plan. Regular practice, continuous learning and adapting to individual patient needs are essential components of becoming proficient in conducting physical examinations. Table 3.6 provides an overview of the physical examination specifically related to pyelonephritis.

Throughout the examination, consider patient comfort and if needs be, modify the examination to accommodate the patient's needs, particularly if they are experiencing pain or discomfort in the flank region. Explain each step of the examination to the patient, addressing any concerns they may have and ensuring their active participation in the process. Hands must be washed before and after the examination.

Examination findings, along with the outcome of the patient's history, can determine if further diagnostic tests or imaging studies are warranted. By focusing on the specific symptoms and signs associated with pyelonephritis, the physical examination contributes to a more accurate diagnosis and guides the subsequent management and treatment of the patient. It is crucial to integrate the physical examination with other diagnostic tools for a comprehensive assessment.

CLINICAL INVESTIGATIONS

Clinical investigations play a crucial role in confirming the diagnosis of pyelonephritis and determining the severity of the condition. The following discussion outlines common clinical investigations that may be used in the diagnostic process (see Table 3.7).

Table 3.6   The physical examination related to pyelonephritis

General inspection	Wash hands prior to performing the physical examination. Observe the patient for signs of distress, fever and general appearance. Pay attention to vital signs, including an elevated temperature, which is common in pyelonephritis. An elevated heart rate and respiratory rate can accompany systemic responses to infection.
Abdominal examination	• Palpate the abdomen for tenderness, particularly over the costovertebral angle (CVA) on both sides. • Ask the patient to lie down comfortably on the examination table. Ideally, in a supine position with the abdomen exposed. • Ask the patient to expose the abdomen, ensuring clear visibility and access to the entire abdominal area. • Locate the CVA. The CVA is the angle formed by the 12th rib and the vertebral column on both sides of the spine. To locate it, first, identify the lower border of the 12th rib, usually at the level of the lumbosacral junction. Then, trace along the spine laterally until you reach the bony prominence of the vertebral column. The point where these two lines intersect is the CVA. • Before palpation, ensure your hands are warm. • Palpate the CVA by placing your hands on the abdomen, with your fingers pointing towards the patient's head. Place the ulnar border (the little finger side) of your hands on the CVA on both sides. • Apply gentle but firm pressure with your fingertips or the base of your hands over each CVA. Start by pressing down with light pressure and gradually increase it to assess for tenderness. • While palpating, ask the patient if they experience any pain or tenderness. Note their response and the specific location of any discomfort. • After assessing the CVA, continue with a more general abdominal examination, palpating other areas for tenderness, guarding, abdominal distension, masses or abnormalities.
Flank examination	Focus on the flank region, where the kidneys are located. Assess for flank tenderness by gently palpating the area on both sides. The patient with pyelonephritis often experiences localised pain in the flank region.
Genitourinary examination	Evaluate the genitourinary system, checking for signs of dysuria, urgency or frequency. In some cases, there may be suprapubic tenderness. Assess for any signs of blood in the urine (haematuria).
Skin examination	Inspect the skin for signs of dehydration, such as dry mucous membranes or poor skin turgor. Dehydration can be a secondary effect of pyelonephritis.
Pelvic examination	For female patients, a pelvic examination may be warranted to rule out gynaecological issues that could mimic symptoms of pyelonephritis.
Neurological examination	In certain cases, particularly in severe or complicated pyelonephritis, assess for signs of altered mental status or confusion, which may be indicative of systemic involvement.

Source: Adapted from European Association of Urology (2023); Hudson and Mortimore (2020).

Table 3.7 Investigations used in the diagnosis of pyelonephritis

Investigation	Discussion
Urinalysis	Urinalysis is a fundamental test for assessing kidney function and identifying signs of infection. Pyelonephritis often presents with pyuria, bacteriuria and haematuria.
Urine culture and sensitivity	A urine culture is performed to identify the specific bacteria causing the infection and determine its susceptibility to antibiotics. This information is crucial for selecting the most effective antibiotic for treatment.
Blood cultures	Blood cultures may be obtained to identify bacteraemia, which can occur in severe cases of pyelonephritis or in individuals with compromised immune systems.
Full blood count	A full blood count helps assess the overall health of the patient and can indicate the presence of infection. Elevated white blood cell count (leucocytosis) is a common finding in pyelonephritis.
Blood urea nitrogen and creatinine	These tests measure kidney function. Elevated levels may indicate impaired renal function, which can be seen in severe cases of pyelonephritis or in the presence of complications such as renal abscess or sepsis.
Inflammatory markers	Measurement of inflammatory markers such as C-reactive protein and erythrocyte sedimentation rate can help gauge the severity of the inflammatory response in the body.
Imaging studies	Ultrasound: ultrasound imaging of the kidneys can identify structural abnormalities, such as kidney stones or abscesses. Computed tomography scan: It provides detailed images of the kidneys and surrounding structures, aiding in the diagnosis of pyelonephritis and identifying complications.

Source: Adapted from European Association of Urology (2023).

The choice and timing of the clinical investigations noted in Table 3.7 will vary depending on individual patient needs.

MANAGEMENT

The management of pyelonephritis depends on the severity of the infection and the underlying cause. The mainstay of treatment is antibiotics, which are usually given orally for mild-to-moderate cases and in severe cases, they are administered intravenously. The choice and duration of antibiotics depend on the type of bacteria, the sensitivity to the drugs and the response to the treatment. The antibiotics should be started as soon as possible after the diagnosis and continued until the infection is cleared. The culture and sensitivity results should be reviewed when they become available and the antibiotics should be changed if necessary. The National Institute for Health and Care Excellence ([NICE] 2018) has provided guidance on antimicrobial use in pyelonephritis; see also European Association of Urology (2023). In addition to antibiotics, analgesia, fluids and rest may be advised to help relieve the symptoms and prevent dehydration (Hudson and Mortimore 2020). In some cases, surgery may be needed to remove any obstructions or drain any abscesses in the kidneys or the urinary tract.

Table 3.8 provides an overview of treatment related to pyelonephritis.

Table 3.8 An overview of treatment related to pyelonephritis

Antimicrobial therapy	Empirical antibiotics: Empirical antibiotic therapy (this refers to the choice of antibiotic based on clinical judgement) is often commenced before the identification of the specific bacteria causing the infection. Commonly prescribed antibiotics include ciprofloxacin, ceftriaxone or trimethoprim-sulfamethoxazole. Culture-directed therapy: Once the results of urine culture and sensitivity tests are available, the antibiotic regimen may be adjusted based on the specific bacteria and its susceptibility to antibiotics. Duration of treatment: Usually, antibiotic treatment lasts for a course of 7–14 days, depending on the severity of the infection and the patient's response to therapy.
Supportive measures	Hydration: Adequate hydration is essential to help flush bacteria out of the urinary tract. Encouraging the patient to drink plenty of fluids is crucial. Pain management: Analgesics such as paracetamol or non-steroidal anti-inflammatory drugs may be used to manage pain and reduce fever. Antipyretics: Medications such as paracetamol can help reduce fever and alleviate associated symptoms. Intravenous antibiotics: For patients with severe symptoms, complications or those unable to tolerate oral medications.
Monitoring	Close monitoring of vital signs, renal function and response to treatment is essential.
Follow-up	Repeat imaging (if required): Follow-up imaging studies, such as ultrasound or Computed tomography scans, may be performed in some cases to assess treatment response and identify any persistent issues. Repeat urine cultures: For some patients, follow-up urine cultures may be conducted to confirm the resolution of the infection and ensure that the prescribed antibiotics were effective.
Prevention of recurrence	Identification and management of risk factors: Addressing and managing underlying risk factors, for example, urinary tract abnormalities or immune system disorders, is key to preventing recurrent infections. In selected cases, prophylactic antibiotics: For individuals with recurrent pyelonephritis, prophylactic low-dose antibiotics may be considered under the guidance of a healthcare provider.

(Continued)

Table 3.8 (*Continued*)

Important considerations	Pregnancy:
	Treatment considerations may differ for pregnant women and antibiotic choices need to be carefully selected to ensure safety for both the mother and the fetus.
	In cases of complicated or recurrent pyelonephritis, consultation with specialists such as urologists or infectious disease specialists may be beneficial.

Source: Adapted from Hewitt et al. (2017), Hudson and Mortimore (2020), NICE (2018) and European Association of Urology (2023).

PROGNOSIS

Pyelonephritis is a common but serious condition that can affect the kidneys and the person's overall health. It can have serious complications if left untreated. Acute pyelonephritis usually responds well to antibiotic therapy; however, this is primarily dependent on the initial severity of the disease. In most cases where there is prompt diagnosis and appropriate treatment, a complete and uncomplicated recovery can be expected within days to weeks. The prognosis is less promising for older people and those with complicating factors or underlying renal disease.

The treatment for pyelonephritis typically involves a combination of antimicrobial therapy, supportive measures and in some cases, there may be a need for hospitalisation for more severe cases. The choice of treatment depends on factors such as the severity of symptoms, the identified causative agent and the presence of complications.

HEALTH TEACHING

Health teaching for patients with pyelonephritis is an important aspect of their care, as it can empower them to understand their condition, take part in their recovery and prevent future episodes. Patients should be informed about the nature of pyelonephritis. Highlight common causes and risk factors, including anatomical abnormalities, kidney stones and factors contributing to recurrent infections.

Being aware of and recognising symptoms, for example, pyrexia, flank pain, urinary urgency, frequency and dysuria, is essential, encouraging patients to seek prompt medical attention if these symptoms arise. Emphasis is placed on ensuring the patient completes the full course of prescribed antibiotics, even if their symptoms improve prior to completion, as this can help to prevent recurrence and antibiotic resistance.

Hydration plays a vital role in flushing bacteria from the urinary tract and supporting kidney function. Patients should be advised to maintain adequate fluid intake and to avoid urinary tract irritants such as caffeine and alcohol.

Good genitourinary hygiene practices, including wiping from front to back after using the toilet, helps prevent the introduction of bacteria into the urethra. Patients should be advised to empty their bladder regularly and completely and to avoid prolonged periods of urine retention.

Preventive measures, such as staying hydrated, urinating after sexual intercourse and using condoms during sexual intercourse, help prevent the transmission of bacteria from the genitals to the urinary tract; wearing breathable underwear may help in reducing the risk of urinary tract infections (Hudson and Mortimore 2020). Avoid the use of spermicides, diaphragms or tampons, as these can irritate the urinary tract and increase the risk of infection.

If kidney stones contribute to the condition, then lifestyle modifications and dietary changes might be recommended. Patients need to understand when to seek immediate medical attention, particularly if symptoms worsen or there are new complications that arise.

Explain to patients about medication management, including the provision of instructions for any additional medications or prophylactic antibiotics. Specific considerations may be needed for pregnant patients.

For some patients, follow-up appointments with healthcare providers are crucial for monitoring recovery. Engaging in open communication, acknowledging the psychosocial impact of recurrent infections and suggesting resources for emotional support are key elements of the health education process. Health teaching should be tailored to meet individual patient's needs, which takes into account their understanding, cultural background and any specific considerations that are related to their health.

CONCLUSION

This chapter has provided a comprehensive understanding of pyelonephritis, a significant renal condition. From its pathophysiology to clinical presentations, risk factors and diagnostic approaches, this condition has been explored. Recognising the importance of early detection and timely intervention is important; clinical investigations, history taking and physical examination are central in order to make an accurate diagnosis.

Understanding the risk factors associated with pyelonephritis is crucial so that health promotion and preventative strategies can be tailored to meet individual needs and enhance care outcomes. Empowering individuals with knowledge about their condition, treatment options and preventive measures that can be taken assists with effective management and encourages a sense of involvement in their own well-being.

GLOSSARY OF TERMS

Abscess: A collection of pus that can occur within the kidney as a complication of severe pyelonephritis.

Bacteraemia: The presence of bacteria in the bloodstream, which can occur in severe cases of pyelonephritis.

Bacteriuria: The presence of bacteria in the urine.

Costovertebral angle: The angle formed by the 12th rib and the vertebral column, often assessed during a physical examination to check for kidney tenderness.

Dysuria: Pain or discomfort during urination, a common symptom of pyelonephritis.

Empirical treatment: Initial treatment based on clinical judgement before the specific causative agent is identified.

Flank pain: Pain or discomfort in the side or back, over the area where the kidneys are located, often associated with pyelonephritis.

Haematuria: The presence of blood in the urine, a common finding in pyelonephritis.

Inflammatory markers: Biomarkers such as C-reactive protein and erythrocyte sedimentation rate indicate the presence of inflammation in the body.

Pyelonephritis: Inflammation of the kidney tissue, usually caused by a bacterial infection.

Pyuria: A term used to describe the presence of an abnormally high number of white blood cells (pus) in the urine.

Renal function: The ability of the kidneys to filter and excrete waste products from the blood.

Sepsis: A severe, life-threatening response to infection that can occur in cases of severe pyelonephritis.

Suprapubic tenderness: Tenderness in the lower abdomen above the pubic bone, associated with urinary issues.

Ultrasound: An imaging technique that uses sound waves to create visual images of the kidneys and urinary tract.

Urinalysis: Laboratory examination of urine to assess for signs of infection, such as pyuria and bacteriuria.

Vesicoureteral reflux: The abnormal flow of urine from the bladder back into the ureters, potentially contributing to pyelonephritis.

White blood cells: Cells of the immune system involved in fighting infection, often elevated in the urine (pyuria) in pyelonephritis.

MULTIPLE CHOICE QUESTIONS

1. What is pyelonephritis?
   a) Inflammation of the bladder
   b) Inflammation of the kidney
   c) Inflammation of the urethra
   d) Inflammation of the prostate

2. Which bacterium is most commonly associated with pyelonephritis?
   a) *Streptococcus*
   b) *Escherichia coli (E. coli)*
   c) *Staphylococcus aureus*
   d) *Pseudomonas aeruginosa*

3. What is a common symptom of pyelonephritis?
   a) Jaundice
   b) Flank pain
   c) Chest pain
   d) Vision changes

4. How is pyelonephritis diagnosed?
   a) Blood pressure measurement
   b) Electrocardiogram (ECG)
   c) Urinalysis and imaging studies
   d) Blood glucose test

5. What risk factor increases the likelihood of developing pyelonephritis?
   a) Regular exercise
   b) Adequate hydration
   c) Diabetes mellitus
   d) Vegetarian diet

6. Where is the costovertebral angle located?
   a) Abdomen
   b) Lower back
   c) Neck
   d) Chest

7. What imaging study is commonly used to assess kidney function and detect abnormalities in pyelonephritis?
   a) Chest X-ray
   b) Magnetic resonance imaging (MRI)
   c) Computed tomography (CT) scan
   d) Ultrasound

8. What is the role of NSAIDs in pyelonephritis management?
   a) Antibiotic treatment
   b) Pain relief and inflammation reduction
   c) Antifungal therapy
   d) Immune system stimulation

9. Which laboratory test assesses the rate at which red blood cells settle and is often used as a marker of inflammation in pyelonephritis?
   a) Complete blood count (CBC)
   b) Erythrocyte sedimentation rate (ESR)
   c) Blood urea nitrogen (BUN)
   d) Procalcitonin

10. What psychosocial aspect should be considered in the care of individuals with recurrent pyelonephritis?
    a) Fear of heights
    b) Fear of the dark
    c) Fear of crowds
    d) Impact on emotional well-being

REFERENCES

Bethel, J. (2012). Acute pyelonephritis: risk factors, diagnosis and treatment. *Nursing Standard* 27 (5): 51–58.

British Medical Journal (2023). Acute pyelonephritis. *BMJ Best Practice.* https://bestpractice.bmj.com/topics/en-gb/3000111 (accessed November 2023).

Colgan, R., Williams, M. and Johnson, J.R. (2011). Diagnosis and treatment of acute pyelonephritis in women. *American Family Physician* 184 (5): 519–526.

European Association of Urology (2023). EAU guidelines on urological infections. https://d56boch luxqnz.cloudfront.net/documents/full-guideline/EAU-Guidelines-on-Urological-infections-2023.pdf (accessed November 2023).

Hagler, D., Harding, M.M., Kwong, J. et al. (2023). *Lewis's Medical-Surgical Nursing,* 12e. St Louis: Elsevier.

Hewitt, I.K., Pennesi, M., Morello, W. et al. (2017). Antibiotic prophylaxis for urinary tract infection-related renal scarring: a systematic review. *Pediatrics* 139 (5): e20163145. pii: peds.2016–3145, doi: 10.1542/peds.2016-3145.

Hudson, C. and Mortimore, G. (2020). The diagnosis and management of a patient with acute pyelonephritis. *British Journal of Nursing* 29 (3): 144–150.

Matuszkiewicz-Rowińska, J., Małyszko, J. and Wieliczko, M. (2015). Urinary tract infections in pregnancy: old and new unresolved diagnostic and therapeutic problems. *Archives of Medical Science* 1 (1): 67–77. doi: 10.5114/aoms.2013.39202

National Institute of Health and Care Excellence (2018). Pyelonephritis (acute): antimicrobial prescribing. https://www.nice.org.uk/guidance/ng111/chapter/Recommendations (accessed November 2023).

UK Health Security Agency (2017). *Klebsiella* species: guidance, data and analysis. https://www.gov.uk/government/collections/klebsiella-species-guidance-data-and-analysis (accessed November 2023).

Acute Kidney Injury CHAPTER 4

Acute kidney injury (AKI) is a sudden and rapid decline in the function of the kidneys. 'Acute kidney injury' has replaced the concept of 'acute renal failure', as it more accurately describes a wide range of injuries (structural damage) and impairment (loss of function) that causes earlier stages of kidney damage, not just kidney failure (Think Kidneys 2018).

Acute kidney injury is characterised by a sudden decline in the kidney's ability to filter and eliminate waste products and excess fluids from the blood. AKI can lead to the accumulation of waste products and electrolytes in the body, which makes it hard for the kidneys to maintain the right balance of fluid and this results in various complications. These life-threatening biochemical disturbances can become a medical emergency. AKI is a term that covers a spectrum of injury to the kidneys that can result from a number of causes which may co-exist. It is a clinical syndrome rather than a biochemical diagnosis (Think Kidneys 2018; National Institute for Health and Care Excellence [NICE] 2023). AKI is a syndrome characterised by a rapid loss of renal excretory function, typically diagnosed by the accumulation of nitrogenous metabolic end products (urea and creatinine) and/or reduced urine output (Can, Rong, and Lixia 2023). AKI happens within a few hours or a few days. Yaqoob and Ashman (2021) add that this abrupt deterioration in renal function is usually (but not always) reversible over days or weeks. AKI can also affect other organs such as the brain, heart and lungs.

PATHOPHYSIOLOGICAL CHANGES ASSOCIATED WITH ACUTE KIDNEY INJURY

The pathophysiology of AKI involves a complex interplay of various factors that ultimately leads to a rapid decline in kidney function along with dysregulation of extracellular volume electrolytes (this refers to an imbalance or disruption in the normal levels and distribution of fluids [volume] and ions [electrolytes] outside the body's cells). A wide range of aetiologies occurs, essentially due to ischaemia, toxicity and inflammation (Gatehouse and Diamond-Fox 2023). The condition can be broadly categorised into three main phases: initiation, maintenance and recovery.

In all cases of AKI, creatinine and urea build up in the blood over several days, and fluid and electrolyte disorders develop. The most serious of these disorders are hyperkalaemia and fluid overload (possibly causing pulmonary oedema). Phosphate retention leads to hyperphosphataemia. Hypocalcaemia is thought to occur because the impaired kidney no longer produces calcitriol (reducing calcium absorption from the gastrointestinal tract) and as a result of hyperphosphataemia causing calcium phosphate precipitation in the tissues. Acidosis develops because hydrogen ions cannot be excreted. With significant uraemia, coagulation may be impaired and pericarditis may develop. Urine output varies with the type and cause of AKI.

A rise in serum creatinine of at least 50% compared to the patient's normal within the previous seven days, together with a fall in urine output, triggers AKI as a diagnosis. Causes of AKI, include impaired renal blood flow, sepsis or cardiac failure, renal disease itself, glomerular nephritis or acute interstitial nephritis as well as obstruction of urine flow (this may be due to renal stones, bladder blood clots, bladder tumour or other pelvic malignancy). The likelihood increases over age 65 years, with coincidental chronic kidney disease or comorbidities, for example, diabetes or cardiac failure and medications or substances that have the potential to cause damage to the kidneys, leading to kidney toxicity (this is termed nephrotoxic). Intrarenal anti-inflammatory drugs are key contributors to nephrotoxicity (Roberts, Eade, and Francis-Wenger 2023). Understanding renal physiology is essential when comprehending the pathophysiology of AKI. The kidney receives approximately 25% of the total cardiac output, and as a result of this, the kidneys are highly susceptible to ischaemia (Gatehouse and Diamond-Fox 2023).

CHRONIC KIDNEY DISEASE

This is also known as chronic renal disease. Chronic kidney disease causes a reduction in kidney function or structural damage (or both) that has been present for more than three months, with associated health implications (Gatehouse and Diamond-Fox 2023). This long-term condition occurs when the kidneys gradually lose their function over time. Chronic kidney disease is a progressive condition, meaning it tends to worsen over an extended period (see also Chapter 5 of this book).

Chronic kidney disease can develop slowly for many years without any symptoms, as the body is able to compensate for a relatively large reduction in kidney function. The condition is often detected when the kidneys are in the final stage of chronic kidney disease or by chance when blood tests and urinalysis are examined as part of health screening (Swales 2022). Acute tubular necrosis (damage to the tubules of the kidneys, which are essential for filtration and processing of urine) occurs when compensatory mechanisms fail as a result of persistent, prolonged or severe insufficient renal perfusion. O'Callaghan (2016) notes that most AKI results from acute tubular necrosis. The consequences include hypoxaemia, inflammation, cellular injury and organ dysfunction.

CATEGORISATION OF ACUTE KIDNEY INJURY

Categorisation of AKI is determined by the anatomical location of the cause (see Figure 4.1):

- Prerenal

- Intrarenal (also called intrinsic)

- Postrenal

The terms prerenal, intrinsic and postrenal are used to categorise different phases or causes of AKI. In categorising in this way, this helps in understanding the underlying mechanisms and factors leading to kidney dysfunction (see Table 4.1).

Understanding the causes of AKI as identified in Table 4.1 is crucial for healthcare professionals to tailor interventions and treatments specific to the underlying cause of AKI. Early recognition and appropriate management can significantly impact outcomes.

The three phases associated with AKI – initiation, maintenance and recovery – are discussed in Table 4.2.

EPIDEMIOLOGY

Acute kidney injury represents a significant cause of mortality and morbidity, both in and out of the hospital and incurs significant healthcare costs (UK Renal Registry 2020). Makris and Spanou (2016) point out that the reported incidence of AKI varies depending on the definition being used, patient population, clinical setting and geographical area. Employing the Kidney

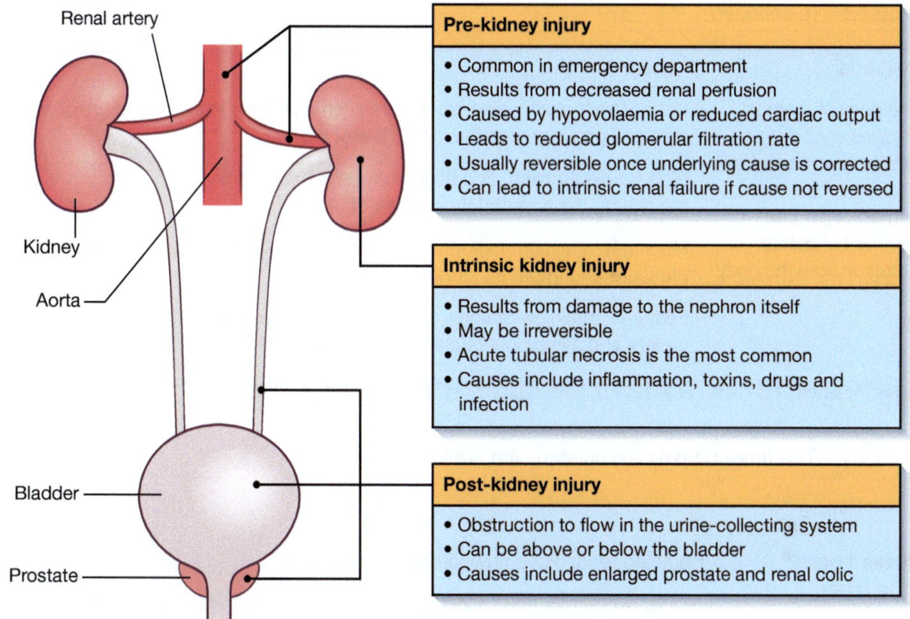

FIGURE 4.1 Categorisation of acute kidney injury

Table 4.1 Causes of acute kidney injury

Prerenal:	Causes include:
Refers to conditions or factors occurring before (or outside) the kidneys, leading to reduced blood flow to the kidneys.	Hypovolaemia: insufficient blood volume due to dehydration, haemorrhage or excessive fluid loss.
	Hypotension: low blood pressure can be caused by conditions such as shock or severe infections.
	Impaired cardiac function: conditions such as heart failure, reduction in blood flow to the kidneys (reduced cardiac output).
	Mechanism: reduced blood flow results in decreased oxygen and nutrient delivery to the kidneys, impacting their ability to function properly.
	Features: the primary characteristic is a decrease in glomerular filtration rate, leading to decreased urine output.
Intrarenal (intrinsic):	Causes include:
This phase involves damage or dysfunction within the kidney tissues themselves.	Acute tubular necrosis: commonly caused by ischaemia or exposure to nephrotoxic substances.
	Glomerulonephritis: inflammation of the glomeruli, kidney's filtering units.
	Interstitial nephritis: inflammation of the kidney's interstitial tissue (a framework of connective tissue and cells surrounding and supporting the nephrons).
	Mechanism: direct damage to kidney structures, particularly the tubules, glomeruli or interstitial tissue, leading to impaired filtration and urine formation.
	Features: manifestations include changes in urine output, electrolyte imbalances and potential signs of systemic illness.

(*Continued*)

Table 4.1 (*Continued*)

Postrenal:	Causes include:
The postrenal phase involves conditions or obstructions occurring after urine has been formed within the kidneys but before excretion from the body.	Urinary tract obstruction: blockages such as kidney stones, tumours or an enlarged prostate preventing the flow of urine. Bladder dysfunction: conditions affecting the bladder, preventing the appropriate release of urine. Mechanism: accumulation of urine within the kidneys, leading to increased pressure and impaired kidney function. Features: backflow of urine, potentially causing dilation of renal pelvis and ureters and symptoms related to obstruction.

Source: Adapted from Hagler et al. (2023), Swales (2022) and Dhaun and Kluth (2023).

Table 4.2 The three phases associated with acute kidney injury

Phase	Discussion
Phase 1: initiation phase	Reduced renal blood flow (prerenal category): • Hypovolaemia (low blood volume) due to dehydration, haemorrhage or other fluid loss. • Decreased cardiac output, as seen in heart failure or shock. • Vasoconstriction of renal blood vessels in response to certain medications or conditions. Direct kidney damage (intrinsic renal category): • Ischaemia: Insufficient blood flow to the kidneys, leading to tissue damage. This can result from prerenal causes or conditions such as sepsis. • Nephrotoxins: Exposure to substances that are toxic to the kidneys, such as certain drugs (e.g. non-steroidal anti-inflammatory drugs some antibiotics), contrast agents or toxins. Urinary tract obstruction (postrenal category): • Blockage of urine flow due to conditions such as kidney stones, tumours or an enlarged prostate.
Phase 2: maintenance phase	Inflammatory response: • Release of inflammatory mediators in response to tissue damage. • Activation of immune cells contributing to inflammation within the kidneys. Cellular injury and dysfunction: • Damage to renal tubular cells, a condition known as acute tubular necrosis, is a common manifestation. • Loss of integrity of the glomerular filtration barrier, leading to impaired filtration function. Oxidative stress: • Accumulation of reactive oxygen species which contributes to cellular damage. • Antioxidant defences may be overwhelmed, further exacerbating tissue injury.

Phase	Discussion
Phase 3: recovery phase	Repair and regeneration: • If the underlying cause is resolved, the kidneys may undergo a repair process. • Regeneration of damaged tubular cells and restoration of normal kidney function. Persistent damage: • In some cases, especially with severe or prolonged injury, there may be persistent damage leading to chronic kidney disease.

Source: Adapted from Hagler et al. (2023); Swales (2022).

Disease: Improving Global Outcomes (KDIGO) definition (Kidney International Supplements 2012), one in five adults and one in three children worldwide experience AKI during a hospital episode of care. Community-acquired AKI on admission to hospital affects approximately 5% in the UK. Severe AKI affects about 130–140 per million population per year. Around 25% of patients with sepsis and 50% of patients with septic shock will have AKI.

The incidence of AKI is increasing. Larmour and Maxwell (2015) and Thornburg and Gray-Vickery (2016) suggest that this is possibly a result of the number of elderly individuals in the population, particularly those with intrarenal conditions and multiple comorbidities. The improved detection of AKI may also have contributed to the rise cited in the literature.

PROGNOSIS

Outcomes vary, and uncomplicated AKI carries a good prognosis, with mortality rates of less than 5–10%. However, AKI complicating nonrenal organ system failure (intensive therapy unit setting) is associated with mortality rates of 50–70%, which have not changed for several decades. Sepsis-related AKI has a significantly worse prognosis than AKI in the absence of sepsis (Yaqoob and Ashman 2021).

RISK FACTORS

Understanding the risk factors for AKI is a key aspect of patient care, as it guides prevention, early detection and enables tailored interventions.

Many people who are at risk of AKI are unaware of the potential causes or what they can do to reduce their risk. When people are aware of the risks and how to maintain their kidney health, then they may be able to prevent AKI. Offering advice to people who are at risk of AKI and their families and carers could reduce the number of people who develop AKI (NICE 2023). Understanding and being aware is vital for improving patient outcomes, making the best use of healthcare resources and promoting kidney health at both individual and population levels (see Table 4.3).

National Institute for Health and Care Excellence (2023) in their quality standard makes known that service providers (for example, GP practices, pharmacies and hospitals) should ensure that processes are in place for those people who are at risk of AKI and their families

Table 4.3 Understanding risk factors for acute kidney injury

Prevention	Knowledge of risk factors allows healthcare providers to identify individuals who may be at a higher risk of developing AKI. This understanding enables preventive measures to be implemented, potentially avoiding or minimising the impact of kidney injury.
Early recognition	Awareness of risk factors helps healthcare professionals recognise the signs and symptoms of AKI promptly. Early detection allows for timely intervention and management, which can improve outcomes and reduce the severity of kidney injury.
Patient education	Informing patients about their specific risk factors empowers them to take proactive measures to reduce their risk. For example, individuals with diabetes may be encouraged to manage their blood glucose levels effectively.
Medication management	Some medications are known to be nephrotoxic. Understanding a patient's medication history allows for the adjustment of medications or for monitoring kidney function more closely to prevent drug-induced AKI.
Tailored monitoring	Patients with known risk factors may require more frequent monitoring of kidney function, such as regular blood tests to assess creatinine levels. This allows healthcare providers to detect any changes early on.
Targeted interventions	Understanding the underlying risk factors helps tailor interventions to address the specific causes of AKI. For example, managing blood pressure in individuals with hypertension or ensuring adequate hydration in those at risk of dehydration.
Research and public health planning	Understanding the epidemiology of AKI risk factors contributes to research efforts and informs public health planning. It helps healthcare systems anticipate challenges and develop strategies to reduce the overall burden of AKI.
Improving overall kidney health	By addressing and managing risk factors, healthcare providers contribute to the overall promotion of kidney health. This is particularly important in preventing the progression of AKI to chronic kidney disease.

and carers if appropriate, to be offered advice on maintaining kidney health. This could be incorporated as part of annual health reviews for people with chronic kidney disease, diabetes, heart failure, dementia, a learning disability or a previous episode of AKI.

An increased risk of AKI is linked to a range of chronic and acute conditions, medicines and social factors. It should be understood that there are some groups of people who are at higher risk of dehydration, some because of their reliance on others to maintain adequate fluid intake. This may include young children, frail older people, people with neurological or cognitive impairment or disability and those with physical disabilities.

People should be offered information that they can easily access and understand themselves or with, if needed support, so that they can communicate and engage effectively with healthcare services. Clear language should be used and the content and delivery of information should be tailored to individual needs and preferences. It should be accessible to people who do not speak or read English, and it should be culturally appropriate. For those who have additional needs related to a disability, impairment or sensory loss, for example, information should be provided as set out in NHS England's Accessible

Table 4.4 Risk factors associated with acute kidney injury

Risk factor	Discussion
Advanced age	The risk of AKI increases with age, particularly in individuals over 65 years.
Previous AKI	Those with previous AKI may have, for example, underlying kidney disease, persistent underlying health conditions, loss of nephrons and incomplete recovery.
Pre-existing chronic kidney disease	Individuals with underlying kidney conditions are more susceptible to developing AKI.
Diabetes	Diabetes, especially if poorly controlled, can contribute to kidney damage and increase the risk of AKI.
Hypertension	Uncontrolled high blood pressure can lead to damage to blood vessels in the kidneys, increasing the risk of AKI.
Cardiovascular diseases	Conditions such as heart failure and myocardial infarction can reduce blood flow to the kidneys, contributing to AKI.
Severe infections	Sepsis can lead to AKI due to the body's inflammatory response and decreased blood flow to the kidneys.
Dehydration	Inadequate fluid intake, excessive fluid loss (e.g., through vomiting, diarrhoea or excessive sweating) or conditions such as heatstroke can contribute to dehydration and increase the risk of AKI.
Use of nephrotoxic medications	Certain medications, such as non-steroidal anti-inflammatory drugs, certain antibiotics and contrast agents used in medical imaging can be nephrotoxic and increase the risk of AKI.
Certain medical procedures	Surgical procedures, especially those involving major blood loss or the use of contrast agents, can pose a risk for AKI.
Urinary tract obstruction	Conditions that obstruct the flow of urine, such as kidney stones or an enlarged prostate, can increase the risk of AKI.
Liver disease	Liver conditions, especially those leading to cirrhosis, can affect metabolism and elimination of waste products, contributing to AKI.
Autoimmune diseases	Conditions such as lupus or vasculitis that involve inflammation of blood vessels can affect kidney function and increase AKI risk.
Certain medications	Some medications may have side effects impacting kidney function, especially if not properly monitored.
Recent major surgery	The stress of major surgery and potential complications can contribute to the risk of AKI.
Malnutrition	Inadequate nutrition or severe malnutrition can compromise overall health, including kidney function.

Source: Adapted from Yaqoob and Ashman (2021), Swales (2022) and Gatehouse and Diamond-Fox (2023).

Information Standard (NHS England 2017) or the equivalent standards for the devolved nations (NICE 2023).

Certain conditions or situations may increase the risk of developing AKI. Those at the highest risk of AKI are identified in Table 4.4.

Having one or more of the risk factors identified in Table 4.4 does not guarantee the development of AKI, and the presence of risk factors does not always lead to kidney injury. However, an awareness of these factors can help to identify individuals who may be at a higher risk and take preventive measures when possible. Early recognition and management of underlying conditions are crucial in reducing the risk of AKI.

CLINICAL PRESENTATION

According to KDIGO (2012), the definition of AKI includes any of the following:

- An increase in serum creatinine by $\geq$25.6 µmol/L within 48 hours.

- An increase in serum creatinine to $\geq$1.5 times baseline, known or presumed within the previous 7 days.

- A urine output of <0.5 mL/kg for 6 hours.

National Institute for Health and Care Excellence (2023) and KDIGO (2012) have produced criteria for the detection of AKI. The criteria include:

- A rise in serum creatinine of 26 µmol/L (0.3 mg/dL) or greater within 48 hours;

- A 50% or greater rise in serum creatinine known or presumed to have occurred within the past 7 days;

- A fall in urine output to <0.5 mL/kg/hour for more than 6 hours in adults and more than 8 hours in children and young people; or

- A 25% or greater fall in estimated glomerular filtration rate (eGFR) in children and young people within the past 7 days.

Acute kidney injury is often categorised into three stages based on the degree of impairment of kidney function: stage 1 (mild), stage 2 (moderate) and stage 3 (severe) (see Table 4.5).

The clinical features of AKI are many, which have been detailed in Figure 4.2.

Box 4.1 discusses the clinical presentation of AKI.

Table 4.5 Stages by specific criteria related to changes in serum creatinine levels and urine output

Stage	Criteria, urine output and clinical features
Stage 1 acute kidney injury (mild)	Criteria: Increase in serum creatinine: 1.5–1.9 times baseline or an absolute increase of 0.3 mg/dL (26.5 µmol/L) or more. Urine output: <0.5 mL/kg/hour for 6 to 12 hours. Clinical features: The earliest stage of AKI. Kidney function is mildly compromised. May not exhibit significant symptoms, but may be subtle changes in urine output or creatinine levels. Early intervention and management can prevent progression to more severe stages.

Stage	Criteria, urine output and clinical features
Stage 2 AKI (moderate)	Criteria:
	Increase in serum creatinine: 2.0–2.9 times baseline.
	Urine output: <0.5 mL/kg/hour for more than 6 to 12 hours.
	Clinical features:
	Moderate impairment of kidney function.
	Greater reduction in urine output.
	Symptoms may become more noticeable, including fatigue, oedema and changes in urine characteristics.
	Increased risk of complications and the need for more aggressive management.
Stage 3 AKI (severe)	Criteria:
	Increase in serum creatinine: 3.0 times baseline or an increase to 4.0 mg/dL (353.6 µmol/L) or more or initiation of renal replacement therapy.
	Urine output: <0.3 mL/kg/hour for 24 hours or anuria for 12 hours.
	Clinical features:
	Severe impairment of kidney function.
	Significant reduction in urine output, which can lead to fluid overload.
	Pronounced symptoms such as severe fatigue, shortness of breath and neurological symptoms.
	Increased risk of complications, including electrolyte imbalances and cardiovascular events.
	Potential need for renal replacement therapy (dialysis) in severe cases.

Source: Adapted from Yaqoob, Ashman, and Feather (2021); KDIGO (2012).

<table><tr><td>BOX 4.1</td><td>

A SUMMARY OF ACUTE KIDNEY INJURY CLINICAL PRESENTATION

</td></tr></table>

- Severity and manifestation are influenced by various factors such as the underlying cause, degree of kidney dysfunction and individual patient characteristics.

- A common feature is a reduction in urine output, which can vary from decreased urine volume (oliguria) to a complete absence of urine (anuria). Important to note that not all cases exhibit decreased urine output. Changes in urine characteristics, including dark colour, foamy consistency or the presence of blood, may be observed.

- Impaired kidney function often leads to fluid retention, resulting in oedema including legs, ankles and face. Additionally, patients may experience fatigue, weakness and an overall sense of malaise due to accumulation of waste products and electrolyte imbalance.

- Respiratory symptoms, such as shortness of breath, may arise from fluid overload affecting the lungs. Electrolyte imbalance can lead to conditions such as hyperkalaemia, hyponatraemia and disturbances in calcium and phosphate levels, causing muscle cramps, weakness and cardiac abnormalities.

(Continued)

BOX 4.1	*(CONTINUED)*

- Gastrointestinal symptoms such as nausea, vomiting and loss of appetite may occur due to the build-up of uraemic toxins. In severe cases, it can impact the central nervous system, resulting in confusion, seizures or coma.

- The risk of bruising, bleeding or an increased bleeding tendency is heightened due to impaired platelet function and coagulation abnormalities.

- Cardiovascular complications, including an increased risk of heart failure, arrhythmias and mortality, are also observed in individuals with AKI.

- The clinical presentation can be further nuanced based on the specific cause:

  - Prerenal (related to decreased blood flow)

  - Intrinsic renal (direct damage to the kidneys)

  - Postrenal (related to urinary tract obstruction)

Source: Dainton (2019).

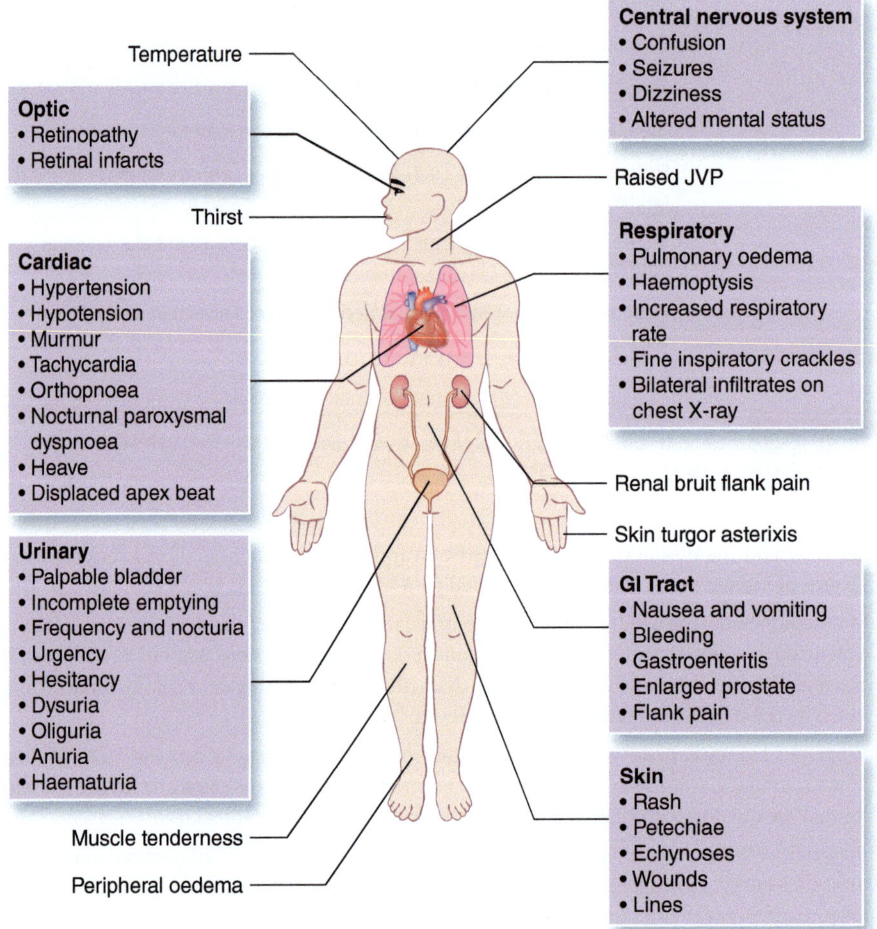

FIGURE 4.2   Clinical features and findings in acute kidney injury

Early recognition, diagnosis and management of AKI are crucial to improving outcomes and preventing further kidney damage. A combination of laboratory tests, imaging studies and clinical assessments is used to effectively diagnose and manage AKI, tailoring interventions based on the severity and underlying causes of the condition.

CLINICAL INVESTIGATIONS AND DIAGNOSIS

Most AKI arises in hospitals as a result of fluid depletion, sepsis or drug toxicity, particularly after surgery, trauma or burns. Usually, there is a fall in urine output and an increase in serum urea and creatinine. A urine output of less than 400 mL per day is called oliguria. It is important to identify those at risk of AKI and take steps to prevent AKI wherever possible (O'Callaghan 2016).

THE HISTORY

The process of history taking plays a critical role in making a diagnosis of AKI. A thorough understanding of the patient's medical history, medications, recent exposures (to infection, toxins, medications) and symptoms is essential for identifying the underlying cause of AKI and informing appropriate management.

A history may indicate the patient has pre-existing renal impairment, hypertension or diabetes mellitus, all of which predispose to renal ischaemia. Passing frank haematuria followed by oliguria is suggestive of glomerulonephritis. In men, urinary frequency, nocturia and a poor stream with hesitancy and dribbling suggest postrenal obstruction that results from prostate disease. Muscle pain and swelling after exercise can suggest rhabdomyolysis. Recent gastroenteritis may indicate *Escherichia coli*-associated haemolytic uraemic syndrome. The person's past medical history may reveal an underlying multisystem disease associated with glomerulonephritis, vascular disease (associated with renal artery stenosis), malignancy (associated with hypercalcaemia) or chronic infection, for example, osteomyelitis or abnormal heart valves that are vulnerable to endocarditis. Assessing the patient's drug history should reveal drug regimens and use of analgesics. Determine if the patient has chronic kidney disease, as this is a risk factor for AKI.

Table 4.6 provides an overview of the history-taking process in relation to AKI.

Undertaking a comprehensive history helps with information gathering regarding the onset, duration and potential causes of AKI. This information, combined with physical examination findings and laboratory tests, contributes to a more accurate diagnosis and guides appropriate management strategies.

Table 4.6   History taking: an overview in relation to acute kidney injury

Onset and duration of symptoms	Inquire about when symptoms started and their progression. This helps determine whether the AKI is acute or chronic and provides a timeline for potential causes.
Urinary symptoms	Explore changes in urinary habits, including alterations in frequency, volume, colour and the presence of blood or foamy urine. Reduced urine output or oliguria is a common feature of AKI.
Medical history	Assess for pre-existing conditions such as diabetes, hypertension, cardiovascular diseases and chronic kidney disease. These conditions can increase the risk of AKI and influence progression.
Medication history	Obtain a detailed list of medications, including prescription and over-the-counter drugs, supplements and herbal remedies. Some medications are nephrotoxic, contributing to AKI.

(Continued)

Table 4.6 (*Continued*)

Recent exposures	Inquire about recent exposures to potential nephrotoxic substances, including contrast agents, certain antibiotics, non-steroidal anti-inflammatory drugs and other toxins. Occupational or environmental exposures should also be considered.
Fluid intake and output	Explore the patient's fluid intake, recent changes in dietary habits and any history of dehydration. Inadequate fluid intake can contribute to prerenal causes of AKI.
Recent infections or illnesses	Ask about recent infections or systemic illnesses; they can trigger inflammatory responses that affect kidney function. Sepsis is a common cause of AKI.
Recent surgeries or procedures	Inquire about recent surgical procedures or medical interventions, especially those involving major blood loss, the use of contrast agents or exposure to nephrotoxic medications.
Pregnancy and gynaecological history	For female patients, consider pregnancy status and any gynaecological issues; certain conditions related to pregnancy or the reproductive system can impact kidney function.
Family history	Explore the family history for any hereditary kidney diseases or conditions that may predispose the patient to kidney-related issues.
Social history	Assess lifestyle factors such as alcohol consumption and recreational drug use. These factors can contribute to dehydration and kidney injury.
Symptoms of systemic illness	Evaluate for symptoms that may suggest the involvement of other organ systems, such as cardiovascular symptoms, neurological changes or respiratory distress.

PHYSICAL EXAMINATION

The physical examination is a key component in making a diagnosis of AKI. While laboratory tests provide essential information about kidney function, the physical examination helps to assess for signs and symptoms that may indicate the presence and severity of AKI, as well as potential underlying causes. The key aspects of the physical examination in relation to diagnosing AKI are outlined in Table 4.7.

Table 4.7 Key aspects of the physical examination related to acute kidney injury

Activity	Rationale	Examination
Blood pressure	Hypertension can be a cause and a consequence of AKI.	Measure blood pressure to assess for hypertension, which can contribute to kidney damage.
Fluid status	AKI can lead to fluid retention or dehydration, affecting overall fluid status.	Assess for signs of fluid overload, such as oedema in the extremities or around the eyes and signs of dehydration, such as decreased skin turgor.
Heart sounds and pulses	Cardiovascular complications can be associated with AKI.	Listen for abnormal heart sounds and assess peripheral pulses. Rapid heart rate or signs of heart failure may indicate cardiovascular involvement.

Activity	Rationale	Examination
Lungs	Pulmonary symptoms may occur in fluid overload.	Auscultate lungs for crackles or signs of pulmonary oedema. Assess respiratory effort and oxygen saturation.
Abdomen	Abdominal pain and tenderness can suggest underlying causes of AKI, such as obstruction or kidney infection.	Palpate the abdomen for tenderness and assess for any abdominal masses. Check for bladder distension, which may be suggestive of urinary outflow obstruction.
Skin	Certain skin findings can be associated with AKI, such as pallor, bruising or a mottled appearance.	Inspect the skin for signs of pallor, ecchymosis (bruising) or livedo reticularis (a mottled purplish discolouration). These findings may suggest underlying vascular involvement.
Neurological examination	Neurological symptoms can occur in severe cases of AKI.	Assess mental status, presence of confusion or neurological deficits. Seizures may occur in cases of severe uraemia.
Extremities	Oedema and changes in skin colour can be signs of fluid imbalance or vascular involvement.	Assess for peripheral oedema, changes in skin colour and temperature. Evaluate pulses and capillary refill.
Genitourinary system	Examination of the genitourinary system can provide clues to the cause of AKI.	Check for suprapubic tenderness, which may indicate bladder distension. Assess for any abnormalities during a genital and pelvic examination.
Signs of systemic illness	AKI can be associated with systemic diseases or infections.	Look for signs of systemic illness, such as fever, which may indicate an underlying infection contributing to AKI.
Monitoring changes over time	Serial examinations help monitor changes in the patient's clinical condition.	Regularly reassess vital signs, fluid status and organ-specific findings to gauge the progression or improvement of AKI.

Assess the fluid volume status. Observe for signs of multisystem disease, cholesterol emboli (which present as a mottled purplish discolouration of the skin), and evidence of intravenous drug use. Muscle swelling or tenderness suggests rhabdomyolysis. The eyes may have hypertensive, diabetic or other diagnostic changes. Examine all wounds, surgical and traumatic, for sepsis. Pulse, blood pressure (lying and standing if necessary), jugular venous pressure and a cardiac examination may indicate volume depletion, cardiac lesions or pericarditis. Examine the chest for pulmonary oedema and evidence of infection or bleeding. Polycystic kidneys may be palpable, and a large palpable bladder can suggest obstruction. Rectal examination may demonstrate prostate or pelvic disease.

The physical examination, combined with clinical history and laboratory investigations, provides an opportunity to identify potential causes, assess the severity of AKI and guide appropriate interventions.

INVESTIGATION AND TESTS

The diagnosis of AKI involves a combination of clinical assessments, laboratory investigations and imaging studies. The goal is to determine the cause, severity and potential complications of AKI. Here are the key investigations and tests commonly used to make a diagnosis of AKI (see also Gatehouse and Diamond-Fox 2023).

LABORATORY TESTS

- Serum creatinine and blood urea nitrogen (BUN):

   These are standard markers of kidney function. An increase in serum creatinine and BUN levels indicates impaired kidney function. Serial measurements help assess the trend over time.

- Estimated glomerular filtration rate:

   Calculated from serum creatinine levels, eGFR provides an estimate of the kidneys' filtering capacity. A lower eGFR indicates reduced kidney function.

URINALYSIS

- Urine output measurement:

   Hourly urine output is monitored to assess for oliguria, a common feature of AKI.

- Urinary measurements/parameters:

   Concentration of urine, urine specific gravity and osmolality can provide insights into the kidneys' ability to concentrate urine.

- Urine dipstick, microscopy and culture should be performed. Leucocytes and nitrites indicate infection. Heavy proteinuria suggests glomerulonephritis or myeloma. Haematuria indicates renal or postrenal disease but may be caused by urinary catheterisation. Myoglobin in the urine suggests rhabdomyolysis and haemoglobin suggests haemolysis. Granular tubular casts may occur in acute tubular necrosis. Red cell casts are diagnostic of glomerular disease. Eosinophils present in urine suggest interstitial nephritis.

ELECTROLYTE AND ACID–BASE BALANCE

- Serum electrolytes (sodium, potassium, calcium, phosphate):

   Imbalances may occur in AKI and monitoring electrolyte levels helps guide treatment.

- Arterial blood gas:

   Assesses acid–base balance, particularly for the presence of metabolic acidosis, a common complication of AKI.

FULL BLOOD COUNT

- Haemoglobin and haematocrit:

   Anaemia may be associated with AKI and these values help assess blood oxygen-carrying capacity.

- Platelet count:

  Thrombocytopenia may occur in severe cases of AKI.

INFLAMMATORY MARKERS

- C-reactive protein and erythrocyte sedimentation rate:

  Elevated levels may indicate an inflammatory component contributing to AKI.

IMAGING STUDIES

- Renal ultrasound:

  Provides information on kidney size, structure and the presence of obstruction or abnormalities.

- CT scan or MRI:

  Useful for evaluating renal anatomy, identifying tumours or assessing for vascular abnormalities.

RENAL BIOPSY

- In certain cases, a renal biopsy may be performed to obtain a tissue sample for detailed examination. Usually reserved for cases where the cause of AKI is unclear or suspected to be related to glomerular or interstitial diseases.

MANAGEMENT

One of the key aims of treatment of AKI is to eliminate the cause, manage the signs and symptoms (clinical features) and prevent complications when the kidneys recover.

Prerenal or postrenal causes must be corrected urgently (O'Callaghan 2016). Life-threatening complications are addressed immediately. This is usually in a critical care unit. Pulmonary oedema is treated with oxygen, intravenous vasodilators, diuretics or dialysis. Dialysis is needed to control hyperkalaemia, pulmonary oedema, metabolic acidosis and uraemic symptoms.

Treatment requires stopping any nephrotoxic drugs and adjusting all drugs excreted by the kidneys, such as digoxin and some antibiotics. Assessing the blood levels of these drugs is helpful.

Management involves addressing the underlying causes, providing supportive care and preventing complications; the approach varies based on the cause, severity of kidney dysfunction and presence of associated complications. Table 4.8 provides the key aspects related to the management of a patient with AKI.

Management of AKI is individualised, which often involves a dynamic and evolving approach based on the patient's response to interventions and the resolution of underlying causes. Timely identification, early intervention and close monitoring are crucial to improving outcomes and preventing further kidney damage. A multidisciplinary team includes nephrologists, intensivists, pharmacists and dietitians to enhance patient care.

Table 4.8 Key aspects related to the management of a patient with acute kidney injury

Fluid management	Volume status assessment:
	Determine the patient's volume status (e.g. dehydration, fluid overload).
	Fluid resuscitation:
	Administer intravenous fluids cautiously to correct dehydration and avoid fluid overload, especially in conditions associated with volume overload or heart failure.
Improve haemodynamic status	Blood pressure control:
	Manage blood pressure to optimise renal perfusion while avoiding excessive pressure that may contribute to further damage.
	Vasopressors:
	In severe cases, vasopressors may be used to maintain adequate perfusion.
Diuretics	Loop diuretics:
	Consider loop diuretics in cases of volume overload; use them with caution to prevent electrolyte imbalances.
Hyperkalaemia	Electrolyte management:
	Address hyperkalaemia promptly with measures such as calcium gluconate, insulin with glucose and potassium-lowering medications.
	Hyponatraemia, hypernatraemia and other electrolyte abnormalities:
	Correct electrolyte imbalances, guided by laboratory values and clinical status.
Nutritional support	Dietary modification:
	Adjust dietary intake based on the patient's nutritional needs and kidney function.
	Restrict potassium, phosphorus and sodium intake as needed.
Avoid nephrotoxic medications	Medication review:
	Review and adjust medications to minimise nephrotoxicity.
	Avoid non-steroidal anti-inflammatory drugs, certain antibiotics and other medications with potential renal side effects.
Renal replacement therapy	Indications:
	Consider renal replacement therapy (dialysis or hemofiltration) in severe cases with life-threatening complications, such as severe fluid overload, hyperkalaemia, metabolic acidosis or uraemic symptoms.
	Initiate renal replacement therapy based on clinical indications and the patient's overall condition.
Monitor and manage complications	Anaemia and erythropoiesis-stimulating agents (a class of medications designed to stimulate the production of red blood cells in bone marrow).
	Address anaemia as needed.

Infection prevention	Monitor for and prevent infections. AKI can compromise the immune system.
Bone health	Manage mineral and bone disorders associated with AKI, including calcium and phosphate balance.
Serial monitoring	Regularly monitor kidney function, electrolytes and other relevant laboratory values.
	Adjust management strategies based on the patient's response and clinical progress.

Source: Adapted from Gatehouse and Diamond-Fox (2023), Swales (2022) and Yaqoob and Ashman (2021).

HEALTH TEACHING

Health teaching for a patient with AKI involves imparting essential information to help the person understand their condition, manage it effectively and prevent complications. Initially, explain to the person that AKI is a sudden decline in kidney function, and explain the common causes such as dehydration, infections and medications. Symptom recognition, encompassing reduced urine output, swelling, fatigue, nausea and confusion, should be emphasised, as well as the importance of seeking prompt medical attention for worsening symptoms.

Medication management is a key aspect, stress adherence to prescribed medications and the avoidance of nephrotoxic substances. Dietary and fluid management guidance includes advising the patient on controlling fluid intake, adhering to dietary restrictions for sodium, potassium and phosphorus and understanding the role of protein in maintaining muscle mass.

Monitoring and self-care education cover vital sign monitoring, regular weighing and the significance of follow-up appointments to adjust treatment plans as needed. Exercise recommendations and lifestyle modifications, such as smoking cessation and limiting alcohol intake (if appropriate), are discussed to promote overall well-being.

Addressing emotional well-being is vital, acknowledging the emotional impact of AKI and providing information on available support services. Involving family members or caregivers in the education process, if appropriate, ensure a comprehensive understanding of the patient's needs. The health education approach is individualised, taking into account the patient's specific needs, understanding and the cultural context, with regular follow-up to provide ongoing support.

CONCLUSION

Acute kidney injury, characterised by a rapid decline in kidney function, demands a comprehensive understanding of its aetiology, clinical presentation and management strategies. The diverse causes of AKI, ranging from prerenal factors such as hypovolaemia to intrinsic renal conditions such as acute tubular necrosis and postrenal obstructions, have been discussed.

Emphasis has been placed on the importance of recognising vulnerable populations. Understanding the phases of AKI, namely prerenal, intrarenal and postrenal, facilitates a targeted approach to diagnosis and management.

The importance of a multidisciplinary approach, involving healthcare providers, family members and support networks, is essential in ensuring comprehensive and patient-centred care. AKI is a dynamic and complex condition that requires not only prompt recognition and intervention but also ongoing monitoring and support.

GLOSSARY OF TERMS

Acute kidney injury (AKI): Sudden and rapid decline in kidney function, characterised by an abrupt decrease in glomerular filtration rate.

Anaemia: A condition characterised by a reduced number of red blood cells or a decrease in haemoglobin levels, common in AKI.

Creatinine: A waste product generated from muscle metabolism; elevated levels may indicate impaired kidney function.

Erythropoiesis-stimulating agents: Medications that stimulate the production of red blood cells, used to manage anaemia in AKI.

Glomerular filtration rate: The rate at which fluid is filtered through the glomerulus, a key measure of kidney function.

Haemodialysis: Renal replacement therapy involving the use of a machine to filter blood when the kidneys are unable to do so.

Intrarenal: Pertaining to conditions or factors that occur within the kidney itself, contributing to AKI.

Loop diuretics: Medications that inhibit sodium reabsorption in the loop of Henle, promoting diuresis.

Nephrotoxic: Substances or medications that are harmful to the kidneys and may contribute to AKI.

Oliguria: Abnormally low urine output, a common symptom of AKI.

Postrenal: Pertaining to conditions or factors occurring after the urine has left the kidney, contributing to AKI.

Prerenal: Pertaining to conditions or factors that occur before the blood reaches the renal tissue, contributing to AKI.

Renal replacement therapy: Therapeutic interventions, such as haemodialysis, are used to replace kidney function in severe cases of AKI.

Uraemia: Build-up of urea and other waste products in the blood, characteristic of advanced kidney dysfunction.

Urine output: The volume of urine produced by the kidneys in a given period, a crucial parameter in AKI assessment.

Vasopressors: Medications that constrict blood vessels, often used to support blood pressure in AKI.

MULTIPLE CHOICE QUESTIONS

1. What is the defining characteristic of acute kidney injury (AKI)?
   a) Gradual decline in kidney function
   b) Sudden and rapid decline in kidney function
   c) Chronic elevation of creatinine levels
   d) Stable glomerular filtration rate (GFR)

2. Which of the following is a common cause of prerenal AKI?
   a) Urinary tract obstruction
   b) Glomerulonephritis

    c) Dehydration
    d) Acute tubular necrosis

3. Intrarenal AKI is associated with damage to which part of the kidney?
   a) Glomerulus
   b) Tubules
   c) Renal pelvis
   d) Ureters

4. Which laboratory test is a reliable indicator of kidney function?
   a) Serum sodium
   b) Blood urea nitrogen (BUN)
   c) White blood cell count
   d) Platelet count

5. What is a common manifestation of oliguria in AKI?
   a) Increased urine output
   b) Dark-coloured urine
   c) Hypotension
   d) Hyperactivity

6. Which class of medications is often associated with nephrotoxicity and can contribute to AKI?
   a) Antibiotics
   b) Antihypertensives
   c) Antipyretics
   d) Anticoagulants

7. Which of the following is a common electrolyte imbalance in AKI?
   a) Hyperkalaemia
   b) Hyponatraemia
   c) Hypocalcaemia
   d) Hypophosphataemia

8. Which of the following is a symptom of uraemia in advanced AKI?
   a) Polyuria
   b) Anaemia
   c) Confusion
   d) Hypertension

9. What is the primary focus of dietary management in AKI?
   a) High protein intake
   b) Sodium and fluid restriction
   c) Potassium supplementation
   d) Phosphorus-rich foods

10. What is the primary reason for monitoring electrolyte levels in AKI?
    a) To assess cardiovascular function
    b) To detect urinary tract infections
    c) To prevent dehydration
    d) To identify and manage imbalances

REFERENCES

Can, W., Rong, L. and Lixia, L. (2023). Incidence and risk factors of acute kidney injury in patients with malignant tumors: a systematic review and meta-analysis. *BMC Cancer* 23. doi: 10.1186/s12885-023-11561-3.

Dainton, M. (2019). Acute kidney injury (Chapter 5). In: *Renal Nursing,* 5e (ed. N. Thomas). Oxford: Wiley.

Dhaun, N. and Kluth, D. (2023). The renal system (Chapter 12). In: *Macleod's Clinical Examination,* 15e (eds. A.R. Dover, A. Innes and K. Fairhurst). London: Elsevier.

Gatehouse, A. and Diamond-Fox, S. (2023). Renal critical care (Chapter 24). In: *Fundamentals of Critical Crae,* (eds. I. Peate, S. Diamond-Fox and B.Hill). Oxford: Wiley.

Hagler, D. Harding., M.M, Kwong., J. et al. (2023). *Lewis's Medical Surgical Nursing,* 12e. St Louis: Elsevier.

Kidney International Supplements (2012). KDIGO clinical practice guideline for acute kidney injury. https://kdigo.org/wp-content/uploads/2016/10/KDIGO-2012-AKI-Guideline-English.pdf (accessed November 2023).

Larmour, K.E. and Maxwell, A.P. (2015). Early intervention can improve outcomes in acute kidney injury. *Practitioner* 259 (1783): 25–28.

Makris, K. and Spanou, L. (2016). Acute kidney injury: definition, pathophysiology and clinical phenotypes. *The Clinical Biochemist Reviews* 37 (2): 85–98.

National Health Service England (2017). Accessible information standard, NHS England. https://www.england.nhs.uk/about/equality/equality-hub/patient-equalities-programme/equality-frameworks-and-information-standards/accessibleinfo/ (accessed November 2023).

National Institute for Health and Care Excellence (2023). Acute kidney injury: prevention, detection and management. https://www.nice.org.uk/guidance/ng148 (accessed November 2023).

O'Callaghan, C. (2016). *The Renal System at a Glance,* 4e. Oxford: Wiley.

Roberts, C., Eade, C. and Francis -Wenger, H. (2023). Diagnostic interpretation (Chapter 11). In: *The Advanced Practitioner* (eds. I. Peate, S. Diamond-Fox and B. Hill). Oxford: Wiley.

Swales, J. (2022). The person with a urinary disorder (Chapter 29). In: *Nursing Practice,* 3e (eds. I. Peate and A. Mitchell). Oxford: Wiley.

Think Kidneys (2018). Acute kidney injury best practice guidance: responding to AKI warning stage test results for adults in primary care (thinkkidneys.nhs.uk). https://www.thinkkidneys.nhs.uk/aki/wp-content/uploads/sites/2/2018/11/Nov18-Primary-Care-Best-Practice-Guidance.pdf. (accessed November 2023).

Thornburg, B. and Gray-Vickrey, P. (2016). Acute kidney injury: limiting the damage. *Nursing* 46 (6): 24–34.

UK Renal Registry (2020). Acute Kidney Injury (AKI) in England – A report on the nationwide collection of AKI warning test scores from 2018. https://ukkidney.org/sites/renal.org/files/AKI_report_FINAL_13072020.pdf (accessed November 2023).

Yaqoob, M.M. and Ashman, N. (2021). Kidney and urinary tract disease (Chapter 36). In: *Kumar and Clark's Clinical Medicine,* 10e. (eds. A. Feather, D. Randall and M. Waterhouse). London: Elsevier.

Chronic Kidney Disease CHAPTER 5

Chronic kidney disease (CKD) is a long-term condition in which the kidneys do not work as well as they should. It describes abnormal kidney function and/or structure. It is a common condition that is often associated with getting older. Over time, CKD can get worse and eventually, the kidneys may cease working altogether, although this is uncommon. Many people with CKD are able to live long lives with the condition. Often, in the early stages, there are no symptoms. It may only be diagnosed if the person has had a blood or urine test for another reason. At a more advanced stage, symptoms can include fatigue, oedematous legs, feet and hands, shortness of breath, nausea and vomiting and haematuria. CKD is usually caused by other conditions that put a strain on the kidneys. Often it is the result of a combination of different problems. CKD can be caused by hypertension, diabetes mellitus, hypercholesterolaemia, kidney infections, glomerulonephritis, polycystic kidney disease, outflow obstruction (such as kidney stones, prostatic enlargement) and long-term regular use of some medicines, for example, lithium and non-steroidal anti-inflammatory drugs (Yaqoob and Ashman 2021; Peate 2022).

The term 'chronic kidney disease' has replaced terms such as chronic renal failure or insufficiency. CKD is a descriptive term and is used for deteriorating kidney function of any underlying cause. CKD implies longstanding (more than three months), potentially progressive impairment in renal function. However, in most cases of stages 1–3 CKD, kidney function does not continue to decline over time and patients do not die as a direct result of kidney disease (Kidney Disease: Improving Global Outcomes [KDIGO] 2013; Yaqoob and Ashman 2021).

ACUTE KIDNEY INJURY

Acute kidney injury (AKI) and CKD differ in several key ways. With AKI, issues relating to the condition occur suddenly and sometimes severely. Often the cause is related to a drug, illness or injury that significantly decreases the function of the kidneys. The symptoms associated with the condition develop suddenly and severely. In most cases, AKI can be reversed once the underlying condition has been treated (see also Chapter 4 of this book).

Table 5.1 provides a comparison between AKI and CKD.

Table 5.1 Comparison between acute kidney injury and chronic kidney disease

Acute kidney injury	Chronic kidney disease
Usually, the condition is caused by medication, illness or injury.	Usually, the condition is caused by a chronic illness.
The decline in kidney function is extreme.	The decline in kidney function is gradual.
The onset of symptoms is sudden and they can be severe.	The symptoms may not appear until the kidney damage is severe.
The treatment is focused on resolving the underlying cause(s).	The treatment is focused on managing the underlying cause.
The condition is most often reversible.	Usually, the condition is non-reversible.

PATHOPHYSIOLOGICAL CHANGES ASSOCIATED WITH CHRONIC KIDNEY DISEASE

Chronic kidney disease is a long-term condition where the kidneys gradually lose their ability to properly filter waste and excess fluids from the blood. Often, this is a result of underlying health issues such as diabetes, hypertension or certain kidney conditions. CKD progresses in stages, with symptoms such as fatigue, swelling and changes in urination patterns becoming more noticeable as kidney function declines. Early detection and management are key components in helping to slow down progression and prevent complications associated with the condition. CKD typically arises from underlying conditions that impose stress on the kidneys. It is often the result of a combination of different problems. CKD may result from any cause of renal dysfunction of sufficient significance:

- Hypertension: over time, this can exert pressure on the tiny blood vessels within the kidneys, leading to impaired kidney function.

- Diabetes: too much glucose in the blood can damage the tiny filters in the kidneys (known as diabetic neuropathy).

- High cholesterol: this can cause a build-up of fatty deposits in the blood vessels that supply the kidneys, which can make it harder for them to work effectively.

- Metabolic syndrome (a cluster of conditions that often occur together, increasing the risk of heart disease, stroke and type 2 diabetes).

- Current or previous history of AKI.

- Kidney infections.

- Glomerulonephritis: inflammation of the glomeruli; these are the small filtering units in the kidneys responsible for removing excess fluids and waste from the blood.

- Autosomal dominant polycystic kidney disease: an inherited condition where growths called cysts develop in the kidneys.

- Obstructions in the urinary flow, such as recurring renal calculi (kidney stones) or prostatic hypertrophy (enlarged prostate), can impede the normal passage of urine.

- Long-term, regular use of certain medicines – such as lithium, non-steroidal anti-inflammatory drugs, some antihypertensive medications and certain antibiotics.

Decreased renal function interferes with the kidneys' ability to maintain fluid and electrolyte homeostasis (electrolyte imbalance occurs). There may be leakage of protein and/or blood into the urine, which results in proteinuria and haematuria (National Institute for Health and Care Excellence [NICE] 2014).

The condition is irreversible and eventually affects all the organs of the body. The parenchyma and the nephrons are destroyed and the renal function progressively diminishes. While there is no cure for the condition, treatment can help delay the progression of CKD, reduce or prevent the development of complications and relieve symptoms. Treatment will depend on how severe the condition is. Management may involve lifestyle changes, medication adjustments and, in some cases, dialysis or kidney transplantation for advanced stages of CKD. Regular monitoring and collaboration with healthcare professionals are crucial for effective management and slowing the progression of the disease.

Table 5.2 provides a discussion of the key aspects of CKD pathogenesis.

Table 5.2 Key aspects of chronic kidney disease (CKD) pathogenesis

Key aspect	Discussion
Hypertension and diabetes	Primary causes:
	Hypertension and diabetes mellitus are two leading causes of CKD.
	Impact:
	Persistent high blood pressure places stress on the blood vessels within the kidneys, diminishing their capacity to filter efficiently.
	Diabetes, especially when uncontrolled, can damage the small blood vessels and nephrons in the kidneys.
Inflammatory processes	Glomerulonephritis: Inflammation of the glomeruli, known as glomerulonephritis, can lead to structural damage and dysfunction in the kidneys.
	Immune system disorders: Conditions where the immune system mistakenly attacks the kidneys, causing inflammation (for example, systemic lupus erythematosus).
Genetic factors	Polycystic kidney disease: Inherited conditions such as polycystic kidney disease result in the formation of fluid-filled cysts in the kidneys, compressing renal tissue and interfering with normal kidney function.
Obstructive processes	Urinary tract obstructions:
	Obstructions in the urinary tract, such as kidney stones or prostatic hypertrophy, can hinder the normal flow of urine, potentially causing harm over an extended period.
Toxic exposures	Certain medications: Long-term use of certain medications, especially those that may be toxic to the kidneys, can contribute to CKD.
Vascular disorders	Atherosclerosis: Hardening and narrowing of the arteries supplying the kidneys can reduce blood flow and contribute to CKD.
Proteinuria	Leakage of proteins: Excessive leakage of proteins into the urine (proteinuria) is a common marker of kidney damage and can be both a cause and a consequence of CKD.
Alterations in the blood flow within the kidneys	Intrarenal haemodynamic changes: Changes in intrarenal haemodynamics can impact the filtration capacity of the kidneys, leading to progressive damage.
Free radicals	Free radicals are reactive molecules that can cause damage to cells. Increased oxidative stress, caused by an imbalance between free radicals and antioxidants, can contribute to kidney cell damage.

Source: Adapted from Swales (2022), Peate (2022) and Gatehouse and Diamond-Fox (2023).

Understanding the diverse factors that contribute to CKD pathogenesis highlights the importance of a comprehensive approach to its management. Early detection, lifestyle modifications and targeted treatments addressing the underlying causes are key in slowing the progression of CKD and preserving kidney function. Regular monitoring plays a key role in the holistic management of CKD.

EPIDEMIOLOGY

The incidence and prevalence of CKD varies depending on the population studied, including ethnic group and socio-economic class. CKD is a common condition; it is present in all adult populations studied. There appears to be some variation in prevalence in different populations; however, the average global prevalence has been reported to be 11.1% in adults. A similar prevalence has been reported in the UK. Based on an estimated global population of 7.6 billion in 2017, this equates to 843.6 million people with CKD stages 1–5; this is approximately double the estimated number with diabetes mellitus (422 million in 2014), (Jager et al. 2019).

RISK FACTORS

It is important to identify factors that are linked with an increased risk of developing CKD so that screening programmes can be targeted at those people who are in high-risk groups.

Chronic kidney disease is influenced by various factors that increase the risk of its occurrence. Lucas and Taal (2023) provide a detailed discussion of the risk factors. Chief among these is uncontrolled diabetes, where elevated blood glucose levels can inflict damage on the kidneys over time. Persistent hypertension, a prevalent risk factor, exerts strain on kidney blood vessels, contributing to kidney damage. Advancing age, with a notable increase in risk after the age of 60 years and a family history of kidney disease, can also increase susceptibility.

Certain ethnicities exhibit an increased predisposition to CKD. Cardiovascular diseases, obesity and smoking are additional risk factors that can adversely impact kidney function. Elevated cholesterol levels contribute to atherosclerosis, further intensifying the risk of kidney damage. Genetic factors, autoimmune diseases such as systemic lupus erythematous and urinary tract infections (UTIs) are also implicated in CKD development.

The prolonged use of certain medications, particularly those with potential kidney toxicity and a history of acute kidney injury, can contribute to the progression of CKD. Additionally, low birth weight and recurrent or severe infections, particularly in the urinary tract or kidneys, can heighten the risk. Recognising these risk factors is pivotal for preventive measures, including regular health assessments, lifestyle modifications and the management of underlying conditions, to mitigate the likelihood of CKD.

CLINICAL PRESENTATION

Often, CKD is usually asymptomatic, and it can go unrecognised as there are no specific symptoms, and it is frequently undiagnosed; when it is diagnosed, it is at an advanced stage, and its clinical presentation can vary among individuals (Centers for Disease Control and Prevention 2023). Most patients with CKD have few symptoms until they have kidney failure. The diagnosis of CKD requires prior biochemical data to allow it to be distinguished from AKI (Dhaun and Kluth 2023).

Since CKD may not present with noticeable symptoms, regular monitoring is crucial for timely detection. As CKD advances, individuals may experience various clinical presentations, including persistent fatigue, weakness and pallor that is related to anaemia, a common complication of kidney disease. An early symptom is uraemic foetor, which is an ammonia odour in the mouth caused by the high concentration of urea in the saliva which subsequently breaks down to ammonia (Gatehouse and Diamond-Fox 2023).

Fluid retention leading to oedema (in the legs, ankles or around the eyes) is another common manifestation. Pulmonary oedema can occur, bringing with it dyspnoea. There can be changes in urinary habits, such as increased or decreased frequency, foamy urine or difficulty urinating.

Hypertension often accompanies CKD and the condition can both contribute to and result from elevated blood pressure. Proteinuria and haematuria are also frequent signs of kidney damage. Electrolyte imbalance occurs, particularly potassium and phosphorus, which can lead to symptoms such as muscle cramps, weakness and bone pain. Anaemia, resulting from reduced production of erythropoietin (a hormone produced by the kidneys and liver), in response to low oxygen levels in the blood, which stimulates the production of red blood cells, erythropoiesis, in the bone marrow, can occur.

CKD also impacts calcium and phosphate metabolism, giving rise to bone and mineral disorders, which may manifest as bone pain and an increased risk of fractures (O'Callaghan 2016). Loss of appetite, nausea and vomiting, unintentional weight loss and pruritus can occur, especially in advanced stages. Cognitive impairment, difficulty concentrating and memory problems may be observed due to the impact of CKD on a person's cognitive function.

It is essential to recognise that the severity and combination of symptoms can vary among individuals (see Figure 5.1).

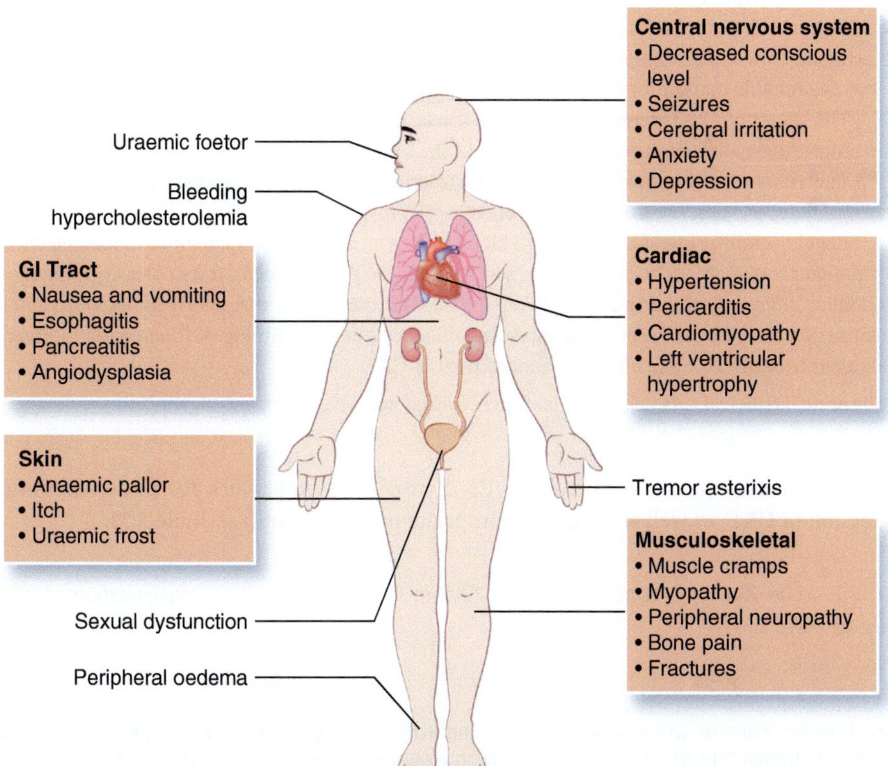

FIGURE 5.1 Clinical manifestations associated with chronic kidney disease

CLINICAL INVESTIGATIONS AND DIAGNOSIS

Making a diagnosis of CKD involves a combination of history taking, physical examination, clinical evaluation, laboratory tests and imaging studies to assess kidney function and identify potential causes. Assessment of renal function allows for the determination of the stage of CKD.

CATEGORISATION OF CHRONIC KIDNEY DISEASE

Chronic kidney disease is usually categorised into five stages based on the level of kidney function, which is measured by the glomerular filtration rate (GFR), (see Box 5.1).

BOX 5.1	GLOMERULAR FILTRATION RATE (GFR) AND ITS ROLE IN THE CLASSIFICATION OF CHRONIC KIDNEY DISEASE

The GFR[a] serves as a crucial measure in assessing kidney function and plays a pivotal role in the classification of CKD into various stages (these are outlined in Table 5.3). GFR essentially quantifies the rate at which the kidneys are able to filter and remove waste products, as well as excess fluid, from the bloodstream. This process occurs in the glomeruli, which are tiny, intricate structures within the kidneys responsible for filtering blood.

GFR is a significant indicator because it reflects the efficiency with which the kidneys perform their essential role in maintaining the body's internal balance. As blood circulates through the kidneys, substances such as creatinine are filtered out and their concentrations are measured to estimate GFR. Creatinine is a by-product of muscle metabolism, and its levels in the blood can serve as a reliable marker for kidney function.

The stages of CKD, determined by GFR levels, offer a systematic framework to understand the progressive nature of kidney dysfunction. In the initial stages, despite signs of kidney damage, GFR may remain within the normal range. However, as the disease advances, GFR declines, indicating a reduced ability of the kidneys to effectively filter blood.

Regular monitoring of GFR through blood tests is an integral component of managing CKD. This ongoing assessment helps gauge the severity of kidney impairment, track disease progression and tailor interventions accordingly. By relying on GFR measurements, informed decisions about treatment strategies can be made, lifestyle modifications and the timing of therapeutic interventions, contributing to more effective management of CKD.

[a] The normal GFR can vary depending on factors such as age, sex, body size and ethnicity.

The five stages used to categorise CKD provide a framework for understanding the progression of CKD as well as guiding appropriate management (see Table 5.3).

Table 5.3 The Kidney Disease: Improving Global Outcomes (KDIGO) classification of chronic kidney disease (CKD)

Stage	Discussion
Stage 1: kidney damage with normal glomerular filtration rate (GFR) (GFR > 90 mL/min):	In this early stage, there is evidence of kidney damage, such as protein or blood in the urine, but the GFR is normal. Kidney function may still be sufficient and individuals may not experience noticeable symptoms.

Stage	Discussion
Stage 2: mildly reduced GFR (GFR 60–89 mL/min):	Kidney function is mildly reduced and individuals may still not exhibit significant symptoms. Kidney damage may progress and it is important to monitor and manage risk factors to slow the advancement of CKD.
Stage 3: moderately reduced GFR (GFR 30–59 mL/min):	This stage is further divided into: • Stage 3a (GFR 45–59 mL/min) • Stage 3b (GFR 30–44 mL/min) Kidney function is moderately reduced and individuals may start experiencing symptoms such as fatigue, fluid retention and changes in urination patterns.
Stage 4: severely reduced GFR (GFR 15–29 mL/min):	Kidney function is significantly diminished and symptoms become more pronounced. Individuals may experience complications such as anaemia, bone disease along with electrolyte imbalances. Preparation for renal replacement therapy, such as dialysis or kidney transplant, becomes important.
Stage 5: kidney failure (GFR <15 mL/min or on dialysis):	This stage is also known as end-stage renal disease, which is marked by a severe decline in kidney function. Symptoms are pronounced and individuals may require renal replacement therapy to sustain life. Options include dialysis or kidney transplantation.

Source: Adapted from KDIGO (2013).

The progression of CKD can vary among individuals and not everyone will go through all the stages cited in Table 5.3. Early detection and management of risk factors, such as diabetes and hypertension, are crucial for slowing the progression of CKD. Regular monitoring of kidney function through blood tests and collaboration with healthcare professionals are essential for individuals at risk or diagnosed with CKD.

MEDICAL HISTORY AND PHYSICAL EXAMINATION

A thorough history helps identify risk factors and potential causes of CKD, such as diabetes, hypertension, family history and exposure to nephrotoxic substances (see Table 5.4). The physical examination can reveal signs of fluid retention, hypertension or other symptoms that are related to kidney disease.

A comprehensive medical history, along with a thorough physical examination and diagnostic tests, forms the basis for the accurate diagnosis and management of CKD. It provides valuable insights into potential causes, risk factors and the overall health of the individual, aiding healthcare professionals in delivering personalised care.

The physical examination for the diagnosis of CKD focuses on assessing signs and symptoms related to kidney dysfunction and its complications. A thorough physical examination, combined with medical history and diagnostic tests, helps to evaluate the patient's overall health and to identify potential causes and manifestations of CKD. See Table 5.5 for an outline of the components related to a physical examination relevant to CKD.

Table 5.4 Components of medical history taking in relation to chronic kidney disease (CKD)

Component	Discussion
Chief complaint and presenting symptoms	Determine why the individual sought medical attention and explore symptoms related to kidney dysfunction, such as fatigue, swelling, changes in urination patterns and discomfort or pain in the kidney area.
Duration and progression of symptoms	Inquire about the duration of symptoms and whether they have progressed gradually or suddenly. This helps assess the chronicity and severity of CKD.
Risk factors	Identify any risk factors that may be associated with CKD, including: • Diabetes: A leading cause of CKD • Hypertension: A major contributor to kidney damage • Family history: Checking for a history of kidney disease among close relatives • Cardiovascular disease: CKD and cardiovascular conditions often coexist
Medication history	Review current and past medications to identify potential nephrotoxic substances or medications that may impact kidney function.
Medical and surgical history	Explore other medical conditions and surgical procedures, as some illnesses and surgeries can contribute to kidney dysfunction.
Dietary habits	Inquire about dietary habits, especially the intake of high-sodium or high-protein foods, as well as fluid intake.
Lifestyle factors	Explore lifestyle factors such as physical activity, smoking and alcohol consumption, which can influence kidney health.
Occupational and environmental exposures	Identify potential exposures to nephrotoxic substances in the workplace or environment.
Urinary symptoms	Ask about changes in urinary habits, including frequency, urgency and the presence of blood or foamy urine.
Fluid retention	Inquire about symptoms of fluid retention, such as swelling in the ankles or around the eyes.
Skin symptoms	Explore any skin-related symptoms, such as pruritis which can be associated with CKD.
Cardiovascular symptoms	Check for symptoms related to cardiovascular complications, such as chest pain or shortness of breath.
Reproductive history	For women, discuss reproductive history, including pregnancy and any complications related to kidney function.
Previous kidney function tests	Review any previous kidney function tests, including serum creatinine and estimated GFR.
Compliance with medications and medical advice	Assess adherence to prescribed medications and medical recommendations.

Table 5.5 An overview of the physical examination required to make a diagnosis of chronic kidney disease (CKD)

Examination component	Discussion
Blood pressure measurement	Hypertension is both a cause and a consequence of CKD. Regular monitoring of blood pressure is essential to assess its impact on kidney function.
Fluid status assessment	Examination for signs of fluid retention, including: Palpation for oedema in the legs. Observation for periorbital oedema (around the eyes). Assessment of jugular venous distension.
Palpation of kidneys	Palpating the abdomen to assess the size, tenderness or abnormalities of the kidneys. Normally, kidneys are not palpable, but enlargement or tenderness may indicate kidney disease.
Cardiovascular examination	Evaluation of the cardiovascular system is crucial due to the association between CKD and cardiovascular disease. This includes: • Auscultation for heart murmurs • Assessment of heart sounds • Examination for signs of heart failure, such as jugular venous distension or peripheral oedema
Respiratory examination	Listening to lung sounds to assess for signs of fluid overload, which can occur in advanced CKD.
Skin examination	Inspection for signs of skin manifestations associated with CKD, such as: • Pallor: Indicative of anaemia • Pruritus: Itching, often related to uraemic toxins in the blood • Ecchymosis: Bruising, which may be related to platelet dysfunction in CKD • Cachexia and signs of malnutrition
Neurological examination	A brief neurological examination may be relevant, especially in advanced CKD, to assess for signs of: • Cognitive impairment • Neuropathy
Musculoskeletal examination	Assessment of signs of bone and mineral disorders associated with CKD, including: • Bone pain • Joint abnormalities
Peripheral vascular examination	Checking pulses and assessing for signs of peripheral vascular disease, which can be associated with CKD.
Reproductive examination (if relevant)	For women, discussing reproductive history, including pregnancies and any complications related to kidney function.
Evaluation of access sites (if on dialysis)	If the patient is on dialysis, inspecting and assessing the vascular access site for signs of infection, clotting or dysfunction.

Source: Adapted from Swales (2022); Hagler, Harding, and Kwong (2023).

INVESTIGATIONS AND TESTS

Diagnosing CKD involves a comprehensive set of investigations to assess kidney function, identify potential causes and monitor associated complications. The diagnostic process will encompass various tests and examinations that are tailored to gain a thorough understanding of the patient's renal health.

There are a number of key blood tests required; for example, the measurement of serum creatinine and blood urea nitrogen may be undertaken. This can offer insight into waste product levels. However, NICE (2021) suggests that serum urea is a poor marker of renal function because it varies significantly with hydration and diet, is not produced constantly and is reabsorbed by the kidney; serum creatinine also has significant limitations. The level can remain within the normal range despite the loss of over 50% of renal function. A full blood count helps assess for anaemia and blood abnormalities associated with CKD.

Determining GFR is a gold standard NICE (2021), a critical measure of kidney function, involving either direct measurement or estimation using specialised equations. Urinalysis, including urine dipstick and microalbumin tests, aids in detecting protein, blood and other indicators of kidney damage.

Imaging studies, for example ultrasound, CT scans or MRIs, offer visual assessments of kidney structure, identifying any structural abnormalities. The patient's medical history and the physical examination contribute valuable information to assess symptoms, risk factors and potential causes of CKD.

Urinalysis, dipstick proteinuria, may suggest glomerular or tubulointerstitial disease. Urine sediment with red blood cells and red blood cell casts suggests proliferative glomerulonephritis. Pyuria and/or white cell casts suggest interstitial nephritis (especially if eosinophils are present in the urine) or UTI.

Proteinuria; the degree of proteinuria correlates with the rate of progression of the underlying kidney disease and is the most reliable prognostic factor in CKD. For the initial detection of proteinuria, a urine sample is used for assessment.

Patients whose initial urinalysis reveals microscopic haematuria should have a urine culture performed to exclude a UTI. Patients over 40 years of age with persistent non-visible/microscopic haematuria in the absence of significant proteinuria or a reduced GFR should be referred to a urology department for further investigation.

An ECG and echocardiography are performed to detect left ventricular hypertrophy, ischaemia and to assess cardiac function.

In some cases, a kidney biopsy may be recommended, especially when the cause of CKD is unclear or when specific indications for tissue examination exist. Monitoring blood electrolyte levels will help to evaluate electrolyte balance and a lipid profile assesses cholesterol and triglyceride levels. This can determine cardiovascular risk, which is elevated in CKD patients.

Additional evaluations include the assessment of parathyroid hormone and vitamin D levels to understand if bone and mineral metabolism have been affected by CKD. Testing for autoimmune disorders may be conducted when autoimmune conditions are suspected.

In certain cases, functional imaging studies, such as a renal scan, may be recommended to evaluate kidney function and blood flow. Plain abdominal X-ray may reveal radio-opaque kidney stones or nephrocalcinosis (a condition characterised by the deposition of calcium salts, such as calcium oxalate or calcium phosphate, in the renal tissue). A CT scan allows for the definition of renal masses and cysts, seen on ultrasound, an MRI for patients who require a CT scan but who cannot receive intravenous contrast.

Regular monitoring for CKD-associated complications involves blood tests to assess anaemia, electrolyte imbalances and bone disorders. The evaluation of cardiovascular risk factors, including blood pressure and lipid levels, provides insights into overall cardiovascular health.

Adopting a comprehensive approach provides an opportunity to determine the severity of CKD, identify underlying causes and develop appropriate management strategies. The specific tests ordered may vary based on individual patient characteristics and clinical findings.

MANAGEMENT

The management of CKD is multifaceted. The focus is to slow the progression of kidney damage, preserve existing renal function, manage complications, provide comfort and improve overall health (Hagler et al. 2023). A primary focus in the management of CKD is the control of underlying conditions contributing to CKD, particularly the management of diabetes and hypertension, as tightly regulated blood glucose and blood pressure can help to slow down kidney deterioration (O'Callaghan 2016).

Medication management plays an important role, with prescription drugs targeting blood pressure control, proteinuria reduction and the management of related complications. This often involves the use of angiotensin-converting enzyme inhibitors or angiotensin II receptor blockers. Dietary adjustments are implemented to alleviate the burden on the kidneys. This includes measures such as sodium restriction, protein intake moderation and management of phosphorus and potassium levels (Swales 2022; Green 2019).

Fluid intake is monitored and managed to prevent fluid overload, especially in the advanced stages of CKD. Medications are regularly assessed and adjusted based on changes in kidney function and overall health, with some medications requiring dose modifications or discontinuation as kidney function declines. Anaemia associated with CKD is addressed through the prescription of medications such as erythropoietin-stimulating agents along with iron supplements (Thomas 2019).

Management input also extends to addressing bone and mineral disorders by monitoring and managing imbalances in calcium, phosphorus and vitamin D. Cardiovascular risk factors are carefully managed, recognising that individuals with CKD are at an increased risk of heart disease. This involves controlling blood pressure, managing cholesterol levels and promoting heart-healthy habits.

Regular monitoring of kidney function through blood tests and other diagnostic studies is fundamental. Management strategies are adjusted based on changes in kidney function and emerging complications are promptly addressed. Education and support are offered to individuals with CKD, ensuring they understand their condition, treatment options and the importance of adherence to medications and lifestyle recommendations (Gatehouse and Diamond-Fox 2023).

In advanced stages of CKD, preparation for renal replacement therapy, such as dialysis or kidney transplantation, is a crucial aspect of management. This involves discussions with the patient and, if appropriate, their family.

Ultimately, individualised care and close collaboration between healthcare providers and individuals with CKD are essential for effective management. Regular monitoring, early intervention and a holistic approach that addresses the root causes and complications of CKD contribute to improved outcomes and a better quality of life for those affected by the condition.

HEALTH TEACHING

Health teaching for individuals with CKD is crucial for empowering them to manage their condition, make informed decisions and maintain overall well-being (Thomas 2019). Table 5.6 outlines some key health teaching requirements for people with CKD.

Table 5.6 Some key health teaching requirements for people with chronic kidney disease (CKD)

Requirement	Discussion
Understanding CKD	Explanation of CKD stages: Provide information about the stages of CKD based on GFR to help individuals understand the progression and severity of their condition.
Management of underlying conditions	Control of diabetes and hypertension: Emphasise the importance of managing diabetes and maintaining glycaemic control and hypertension through medication adherence, regular monitoring and lifestyle modifications.
Medication adherence	Explanation of medications: Provide advice to individuals about their prescribed medications, their purpose, potential side effects and the importance of adherence to the prescribed regimen.
	All prescribed medication is reviewed regularly to ensure appropriate doses.
Dietary modifications	Sodium restriction: Emphasise the need for limiting sodium intake to manage fluid balance and blood pressure.
	Protein restriction: Discuss the importance of moderating protein intake to reduce the workload on the kidneys.
	Phosphorus and potassium management: Provide guidance on managing phosphorus and potassium levels through dietary modifications.
	Consultation with dietitian: Facilitate consultations with a dietitian to create personalised dietary plans.
Fluid intake management	Fluid restriction: If necessary, explain the importance of monitoring and restricting fluid intake to prevent fluid overload.
Lifestyle changes	Regular exercise: Encourage the adoption of regular physical activity tailored to individual abilities.
	Smoking cessation: Emphasise the importance of quitting smoking to reduce cardiovascular risk.
	Maintaining a healthy weight: Provide guidance on achieving and maintaining a healthy weight.
	Lifestyle modifications contribute positively to overall health and may assist in managing underlying conditions.
Symptom recognition	Offer information on warning signs: Teach individuals to recognise and report symptoms such as changes in urination, swelling, fatigue and signs of complications such as anaemia.
Anaemia management	Importance of iron and erythropoietin-stimulating agents medications: Explain the role of iron supplements and erythropoietin-stimulating agents in managing anaemia and improving energy levels.
Bone and mineral disorder education	Calcium and vitamin D: Provide information on maintaining proper calcium and vitamin D levels to support bone health.

Requirement	Discussion
Cardiovascular risk reduction:	Blood pressure and cholesterol management: Reinforce the need for controlling blood pressure and managing cholesterol levels to reduce cardiovascular risk.
Regular monitoring:	Importance of follow-up visits: Stress the importance of regular medical check-ups and laboratory tests to monitor kidney function and overall health.
Emotional support and mental health:	Coping strategies: Discuss coping strategies for dealing with the emotional and psychological aspects of living with a chronic condition.
Preparation for renal replacement therapy	Dialysis and transplant education: If applicable, provide information on the different renal replacement therapy options. These include dialysis and kidney transplantation.
Engagement in decision-making	Shared decision-making: Encourage individuals to actively participate in decisions related to their care and treatment plans.

Source: Adapted from Thomas (2019).

Discuss nephrotoxins such as intravenous radiocontrast agents, non-steroidal anti-inflammatories and aminoglycosides (some antibiotics). Promote immunisation against influenza, pneumococcus and COVID-19. Health teaching must be individualised. It has to take into account the patient's specific needs, preferences and cultural considerations. Regular reinforcement and ongoing support are essential components of effective health education for individuals with CKD.

CONCLUSION

Chronic kidney disease is common, affecting around 11% of the adult population globally, and the prevalence of CKD rises sharply with increasing age. CKD is important as it is associated with several adverse outcomes. These include increased risk of cardiovascular events, AKI and progression to end-stage kidney disease.

Risk factors for CKD include older age, male sex, certain ethnicities, lower socioeconomic status, genetic factors and hypertension. There are several causes of CKD, including diabetes mellitus, glomerulonephritis, genetic disorders, cardiovascular disease, multisystem diseases, drugs, urological conditions, infections and AKI.

It is important to identify individuals at high risk of having CKD in order for them to undergo testing to facilitate early diagnosis and treatment to reduce cardiovascular risk and slow progression towards end-stage kidney disease.

GLOSSARY OF TERMS

Albuminuria: Presence of excess albumin (a protein) in the urine, indicating kidney damage.

Blood urea nitrogen: A blood test measuring the amount of nitrogen in the blood that comes from urea, a waste product.

Creatinine: A waste product from muscle metabolism that is filtered out of the blood by the kidneys. Elevated levels may indicate impaired kidney function.

Chronic kidney disease: A long-term condition where the kidneys are not functioning properly.

Dialysis: A medical procedure used to filter and purify the blood when the kidneys are not functioning adequately.

Erythropoietin: A hormone produced by the kidneys that stimulates the production of red blood cells in the bone marrow.

Glomerular filtration rate: A measure of how efficiently the kidneys filter waste and excess fluid from the blood.

Glomerulonephritis: Inflammation of the glomeruli, the filtering units of the kidneys.

Haemodialysis: A type of dialysis where blood is filtered using a machine outside the body.

Nephrocalcinosis: Deposition of calcium salts in the kidney tissue.

Nephron: The functional unit of the kidney responsible for filtering blood and producing urine.

Parathyroid hormone: Hormone that regulates calcium and phosphorus levels in the blood.

Renal replacement therapy: Treatment methods for replacing the lost kidney function, including dialysis and kidney transplantation.

Serum creatinine: Blood test measuring the level of creatinine in the blood.

Systemic lupus erythematosus: An autoimmune disease that can affect various organs, including the kidneys (lupus nephritis).

Urinalysis: Laboratory examination of urine to detect abnormalities or diseases, including kidney-related conditions.

Vitamin D: Essential for maintaining calcium and phosphorus balance in the body; its activation occurs in the kidneys.

MULTIPLE CHOICE QUESTIONS

1. What is the primary function of the kidneys?
   a) Digestion
   b) Filtration of blood and waste elimination
   c) Respiratory gas exchange
   d) Hormone secretion

2. Which of the following is a common risk factor for CKD?
   a) Regular exercise
   b) Hypertension
   c) Low dietary sodium intake
   d) Adequate fluid intake

3. What does GFR stand for in the context of CKD?
   a) Glomerular filtration rate
   b) Gastrointestinal function ratio
   c) Glycaemic feedback response
   d) Glucose filtration rate

4. How is CKD staged based on GFR levels?
   a) By age
   b) By blood pressure

    c) By cholesterol levels
    d) By the severity of kidney dysfunction

5. What is the primary goal in managing CKD?
    a) Reversal of kidney damage
    b) Prevention of further kidney damage
    c) Restoration of normal kidney function
    d) Elimination of proteinuria

6. Which dietary modification is commonly recommended for individuals with CKD?
    a) High protein intake
    b) Low sodium intake
    c) High potassium intake
    d) Unlimited fluid intake

7. What is the role of erythropoietin in CKD?
    a) Regulates blood pressure
    b) Stimulates red blood cell production
    c) Controls calcium levels
    d) Enhances urine concentration

8. Which of the following is a symptom of advanced CKD?
    a) Polyuria
    b) Haematuria
    c) Fatigue
    d) Hypertension

9. What is the primary source of creatinine in the blood?
    a) Liver
    b) Lungs
    c) Muscle metabolism
    d) Bone marrow

10. What is nephrocalcinosis?
    a) Inflammation of the glomeruli
    b) Build-up of calcium salts in the kidney tissue
    c) Surgical removal of a kidney
    d) Autoimmune disorder affecting the kidneys

REFERENCES

Centers for Disease Control and Prevention (2023). Chronic kidney disease in the United States, 2023. https://www.cdc.gov/kidneydisease/pdf/CKD-Factsheet-H.pdf (accessed November 2023).

Dhaun, N. and Kluth, D. (2023). The renal system (Chapter 12). In: *Macleod's Clinical Examination,* 15e (eds. A.R. Dover, A. Innes and K. Fairhurst). London: Elsevier.

Gatehouse, A. and Diamond-Fox, S. (2023). Renal critical care (Chapter 24). In: *Fundamentals of Critical Crae* (eds. I. Peate, S. Diamond-Fox and B. Hill). Oxford: Wiley.

Green, S. (2019). Nutrition and health (Chapter 21). In: *Alexanders Nursing Practice,* 5e (ed. I. Peate). London: Elsevier.

Hagler, D., Harding, M.M., Kwong, J. et al. (2023). *Lewis's Medical Surgical Nursing,* 12e. St Louis: Elsevier.

Jager, K.J., Kovesdy, C., Langham, R. et al. (2019). A single number for advocacy and communication – worldwide more than 850 million individuals have kidney diseases. *Kidney International* 96 (5): 1048–1050. doi: https://doi.org/10.1016/j.kint.2019.07.012

Kidney Disease: Improving Global Outcomes (KDIGO) (2013). KDIGO 2012 clinical practice guideline for the evaluation and management of chronic kidney disease. *Kidney International Journal Supplementary* 3 (1): 1–150.

Lucas, B. and Taal, M.W. (2023). Epidemiology and causes of chronic kidney disease. *Medicine* 51 (3): 165–169. doi: https://doi.org/10.1016/j.mpmed.2022.12.003

National Institute for Health and Care Excellence (2014). Chronic kidney disease in adults: assessment and management. https://www.nice.org.uk/guidance/cg182 (accessed November 2023).

National Institute for Health and Care Excellence (2021). Chronic kidney disease: assessment and management. https://www.nice.org.uk/guidance/ng203 (accessed November 2023).

O'Callaghan, C. (2016). *The Renal System at a Glance,* 4e. Oxford: Wiley.

Peate, I. (2022). *Anatomy and Physiology for Nursing and Healthcare Students a Glance,* 2e. Oxford: Wiley.

Swales, J. (2022). The person with a urinary disorder (Chapter 29). In: *Nursing Practice,* 3e (ed. I. Peate and A. Mitchell). Oxford: Wiley.

Thomas, N. (2019). Chronic kidney disease (Chapter 6). In: *Renal Nursing,* 5e (ed. N. Thomas). Oxford: Wiley.

Yaqoob, M.M. and Ashman, N. (2021). Kidney and urinary tract disease (Chapter 36). In: *Kumar and Clark's Clinical Medicine,* 10e (eds. A. Feather, D. Randall and M. Waterhouse). London: Elsevier.

Renal Calculi CHAPTER 6

Renal calculi are known by several names, and these terms are often used interchangeably. Some of the common names for renal calculi include:

- Kidney stones: Referring to the fact that the stones are located in the kidneys.

- Nephrolithiasis: The medical term for the condition of having kidney stones. These occur in the upper urinary tract.

- Urolithiasis: A broader term that includes the formation of stones in the urinary tract, encompassing the kidneys, ureters, bladder and urethra.

- Calculi: This term is used to describe abnormal stone-like structures and it is often applied to kidney stones.

- Kidney calculi: Specifically referring to stones in the kidneys.

- Renal calculi: Similar to kidney calculi, this term specifies stones in the renal (kidney) area.

- Ureteral stones: If the stones move from the kidneys into the ureter, they may be called ureteral stones.

- Bladder stones: If stones form in the bladder, they are referred to as bladder stones.

These terms are used in medical contexts to describe various aspects of kidney stone formation, location and related conditions. The choice of terminology often depends on the specific location of the stones within the urinary tract. It is the kidney where most stones originate.

Urinary calculi are solid particles in the urinary system. They may cause pain, nausea, vomiting, haematuria and possibly, chills and fever as a result of secondary infection. Diagnosis is based on radiological imaging, usually a non-contrast computed tomography (CT) scan, the imaging technique of choice.

Humankind has been afflicted by urinary stones for centuries dating back to 4000 BC (Lopez and Hoppe 2008) and it is the most common disease of the urinary tract. The prevention of renal stone recurrence still brings serious health challenges. Kidney stones have been associated with an increased risk of chronic kidney diseases, end-stage renal failure, cardiovascular diseases, diabetes and hypertension. It has been suggested that kidney stones may be a systemic disorder linked to metabolic syndrome (see Box 6.1).

BOX 6.1   METABOLIC SYNDROME

Metabolic syndrome is a cluster of conditions that occur together, increasing the risk of heart disease, stroke and type 2 diabetes.

Metabolic syndrome is also known as:

- Syndrome X: This term was initially used to describe metabolic syndrome and its association with insulin resistance. The 'X' signifies the unknown factor linking the various components of the syndrome.

(Continued)

BOX 6.1	*(CONTINUED)*

- Insulin resistance syndrome: This name highlights one of the key underlying factors of metabolic syndrome, where the body's cells become less responsive to insulin, leading to elevated blood sugar levels.

- Dysmetabolic syndrome: This term emphasises the abnormal metabolic processes associated with the syndrome.

Source: Adapted from Hagler et al. (2023).

PATHOPHYSIOLOGICAL CHANGES ASSOCIATED WITH RENAL CALCULI

The pathophysiology of renal calculi involves a complex interaction of factors. The formation of renal calculi comprises various stages within the urinary system. These various interactions lead to the formation of solid crystalline structures (which are tiny, solid formations made up of specific substances such as calcium, oxalate, phosphate, uric acid or cystine) within the urinary system, primarily in the kidneys.

The formation of renal calculi begins with an imbalance in the concentration of minerals in the urine, which is a condition that is known as supersaturation. This occurs when minerals surpass their solubility limits. Influencing factors include diet, fluid intake and genetic predisposition.

Within a supersaturated environment, crystals can form. These act as nuclei for the development of larger solid structures. These crystals may aggregate and adhere to the surfaces of renal tubules or other kidney structures. The aggregation process is influenced by urinary pH (the measurement of the acidity or alkalinity of urine), the presence of substances that promote or inhibit crystal formation and urine flow characteristics. The outcome is the formation of kidney stones, with types such as:

- Calcium oxalate

- Calcium phosphate

- Uric acid

- Cystine stones

See Table 6.1 for a discussion regarding the type of stones.

Table 6.1 Kidney stones

Type of stone	Discussion
Calcium stones	These are the most common types of stones and contain calcium oxalate, calcium phosphate or both.
	Certain factors increase the likelihood of developing kidney stones, including low urine volume, high levels of calcium and oxalate in urine and low levels of citrate. About 65% of people with kidney stones have a condition called hypercalciuria, where there is too much calcium in the urine. Often, the cause is unknown (idiopathic), but it can be linked to increased absorption of calcium in the intestines, obesity and high blood pressure.
	It is advised to drink more fluids and cut back on calcium, sodium and animal protein in the diet. Having too much calcium in the body, often from excessive intake or conditions such as hyperparathyroidism, can lead to hypercalciuria.

Type of stone	Discussion
	Consuming too much dietary sodium increases urine calcium levels, while animal protein intake also raises these levels. Oxalate, a substance in urine, can increase due to a high dietary intake, absorption issues in the colon or certain genetic conditions. Low citrate levels in urine, known as hypocitraturia, can be due to unknown causes or conditions such as distal renal tubular acidosis, affecting citrate metabolism.
Urate stones	Urate stones, also known as uric acid stones, are a type of kidney stone that forms due to the accumulation of uric acid in the urine. Uric acid is a natural by-product of the breakdown of purines, which are substances found in certain foods and body tissues. When the concentration of uric acid in the urine becomes too high, it can crystallise and form solid stones.
Cystine stones	These stones form due to the build-up of the amino acid cystine in the urine. Cystine is one of the building blocks of proteins, and under normal circumstances, it should be adequately reabsorbed by the kidneys. However, individuals with a genetic condition known as cystinuria experience a defect in a protein (the dibasic amino acid transporter), leading to reduced reabsorption of cystine in the renal tubules.
Infection stones	Infection stones, also known as struvite or magnesium ammonium phosphate stones, are a type of kidney stone that forms as a result of certain bacterial infections in the urinary tract.
	These are often large staghorn calculi containing magnesium, ammonium, phosphate and calcium phosphate. Infection, usually with *Proteus* species, produces urease, which splits urea to produce ammonium ions.
	The rise in pH promotes calcium phosphate crystallisation and the ammonium crystallises with magnesium and phosphate.

Source: Adapted from O'Callaghan (2016), Shaw (2016) and Yaqoob and Ashman (2021)

As the kidney stones grow or as they move within the urinary tract, they can cause obstruction, which can cause urinary stasis and impaired drainage. Obstruction increases the risk of bacterial infection and the presence of stones will trigger an inflammatory response, which will contribute to symptoms and potential complications.

Symptoms of renal calculi, including severe pain (renal colic), haematuria, dysuria (painful urination) and frequent urination, often result from stone movement and the irritation that is caused. Alelign and Petros (2019) note that a kidney stone will not cause any symptoms until it begins to pass through the urinary tract. The central feature is pain, which occurs as the stone moves through the kidney or through any part of the urinary tract, including the ureters, bladder and urethra. Complications can include urinary tract infections and hydronephrosis, where the kidneys swell due to impaired drainage.

Various factors contribute to renal calculi formation, such as genetic predisposition, dietary choices, dehydration and specific medical conditions, for example, hypercalciuria (higher than normal levels of calcium in the urine) or hyperoxaluria (higher than normal levels of oxalate in the urine). Understanding these factors is crucial for developing preventive measures and management strategies. Lifestyle modifications, including dietary changes and increased fluid intake, play a vital role. Medical interventions, such as medications to modify urine composition or procedures to break down or remove stones, may be necessary based on the type and size of the stones.

TYPICAL SITES OF STONE IMPACTION IN THE URINARY TRACT

- Ureteropelvic junction

- Distal ureter (at the level of the iliac vessels)

- Ureterovesical junction

Larger calculi are more likely to become lodged. Typically, a calculus must have a diameter >5 mm to become lodged. Calculi ≤5 mm are more likely to pass spontaneously. Calculi vary from microscopic crystalline foci to calculi several centimetres in diameter. A large calculus, known as a staghorn calculus, can fill an entire renal calyceal system (see Figure 6.1). Figure 6.2 shows the position of a lower pole stone. The formation of stones within the urinary tract is called urolithiasis (Swales 2022). European Association of Urology ([EAU] 2023) states

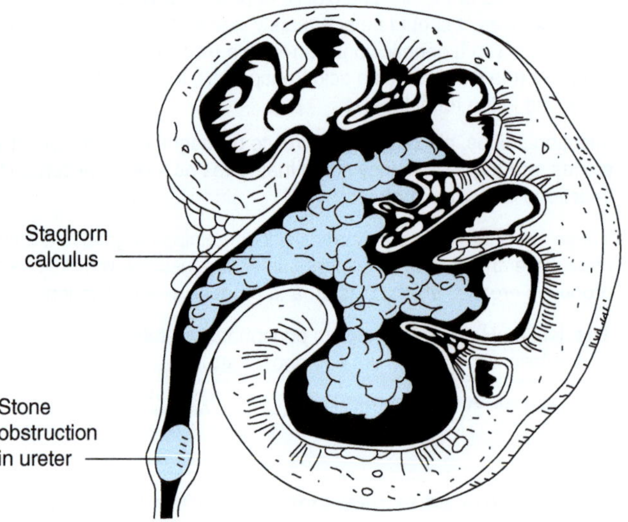

FIGURE 6.1   Renal calculi

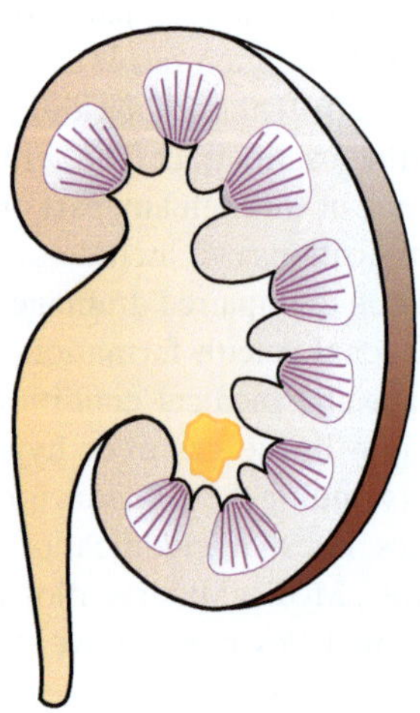

FIGURE 6.2   A lower pole stone

that stones can be classified according to their anatomical position: upper, middle or lower calyx; renal pelvis; upper, middle or distal ureter; and urinary bladder.

A comprehensive understanding of the pathophysiology of renal calculi allows for the provision of tailored interventions that would be required to address specific contributing factors and to provide effective care for individuals affected by this condition.

The composition of kidney stones can vary among individuals and a person may have a combination of different types.

Table 6.2 provides an overview of the pathogenesis of renal calculi.

Table 6.2 An overview of the pathogenesis of renal calculi

Sequence of events	Discussion
Supersaturation of urine	The process often begins with the supersaturation of urine, where the concentration of certain substances such as calcium, oxalate, phosphate or uric acid exceeds their solubility limits. This can result from various factors, including dietary habits, fluid intake and underlying medical conditions.
Nucleation	Supersaturated urine provides an environment that is conducive to nucleation, the initial formation of tiny crystals. These crystals can be composed of calcium oxalate, calcium phosphate, uric acid or other substances, depending on the specific composition of the urine.
Crystal growth	Once nucleation occurs, crystals can grow in size. Factors such as the concentration of stone-forming substances, urine pH and the presence of inhibitors or promoters influence the growth of these crystals.
Crystal aggregation	Crystals may aggregate, forming larger structures. This process can be influenced by the interaction of crystals with cells or matrix components in the kidneys.
Stone formation	As crystals continue to aggregate and adhere to the inner surfaces of the kidney, stones begin to take shape. The stones can vary in size and composition, depending on the types and amounts of substances involved.
Promoters and inhibitors	The presence of substances that promote or inhibit stone formation plays a crucial role. Inhibitors, such as citrate and magnesium, can help prevent crystal aggregation, while promoters, such as high levels of oxalate or low urine volume, contribute to stone formation.
Genetic factors	Genetic predisposition can influence an individual's susceptibility to kidney stones. Genetic disorders, such as cystinuria or hyperoxaluria, can increase the risk of specific types of stones.
Metabolic abnormalities	Certain metabolic abnormalities, such as hypercalciuria, hyperuricosuria or hyperoxaluria, can disrupt the balance of minerals in the urine, contributing to stone formation.
Dehydration	Inadequate fluid intake leading to concentrated urine can facilitate the crystallisation and aggregation of stone-forming substances.
Underlying medical conditions	Conditions such as urinary tract infections, renal tubular acidosis or other anatomical abnormalities can create an environment conducive to stone formation.

EPIDEMIOLOGY

Dawson and Thomas (2012) note that kidney stone disease typically presents between the ages of 20 and 60 years and is more prevalent in hot climates. It affects about 10% of people over their lifetime, incidence increases with age; 50% will have a recurrence within 5–10 years and 75% within 20 years. Developed countries have seen rapid increases over the last 30 years, especially in women, in whom incidence is now almost equal to that of men. In 2018, Public Health England and Industrial Injuries Advisory Council reported that the annual incidence is one to two cases per 1000 people, with incidence peaking between 40 and 60 years for men and in the late 20s for women. It has been estimated that 12% of men and 6% of women will have one episode of renal colic at some stage in their life. Stones occur more commonly in Caucasian and Asian patients (Shaw 2016). It is likely that dietary factors, such as increased protein consumption and reduced fluid intake, along with sedentary occupations increase the risk of stone disease.

RISK FACTORS

Identifying and addressing risk factors can help to reduce the risk of kidney stone formation and promote kidney health. The risk factors associated with the development of renal calculi are outlined in Table 6.3.

It should be noted that those identified in Table 6.3 are risk factors, not causes. Having one or more of these factors does not necessarily mean that a person will develop kidney stones.

Table 6.3   Risk factors associated with the development of renal calculi

Risk factor	Discussion
Age and gender	The incidence of urinary stones varies by age, with low incidence occurring in children and elderly people. The risk of stone formation is higher in men than in women. Although there is some evidence that the risk is now becoming more equal because of lifestyle factors such as obesity and a Western diet.
Ethnicity	Urinary stones primarily occur in white people, followed by black people and Asian people.
Diet	Excessive dietary intake of oxalate, urate, sodium and animal protein is associated with increased stone formation.
Chronic dehydration	Fluid intake is inversely proportional to the risk of stone formation. A low urine output can produce a higher concentration of urinary solutes, leading to stone formation.
Obesity	The prevalence and incident risk of kidney stones is directly correlated with higher weight and body mass index in both genders.
Environmental factors	High ambient temperatures increase the risk of stone formation due to the impact of temperature on fluid status and urine volume. Chronic lead and cadmium exposure are also associated with stone formation (this occurs in occupations that involve exposure to lead and cadmium and often occurs in industries where these metals are used in various processes).
Family history	A positive family history is associated with an increased risk of forming stones. A person forming a stone is twice as likely as a non-stone former to have a first-degree relative with a history of stones.

Risk factor	Discussion
Medical history	Certain conditions increase the risk of stone formation. These include:
	• Anatomical abnormalities of the urinary tract:
	Horseshoe kidney, ureteral stricture, ureteropelvic junction obstruction, ureterocoele, caliceal diverticulum, caliceal cyst, or tubular ectasia (medullary sponge kidney) and vesico-uretero-renal reflux.
	• Gastrointestinal conditions:
	Jejuno-ileal bypass, intestinal resection, Crohn's disease, malabsorptive conditions, enteric hyperoxaluria after urinary diversion and bariatric surgery.
	• Genetic conditions:
	Cystinuria, primary hyperoxaluria, renal tubular acidosis and cystic fibrosis.
	• Conditions which alter urinary volume, pH, and/or concentrations of certain ions:
	Hyperparathyroidism, nephrocalcinosis, diabetes, hypertension, polycystic kidney disease, gout, sarcoidosis, spinal cord injury, neurogenic bladder and increased levels of vitamin D.
Medications	Certain medicines may increase the risk of developing recurrent kidney stones. These include aspirin, antacids, diuretics, certain antibiotics, certain antiretroviral medicines (used to treat HIV) and certain anti-epileptic medicines.

Source: Adapted from EAU (2023); Shang et al. (2017).

CLINICAL PRESENTATION

The clinical presentation of renal calculi can manifest in various ways. Presentation is influenced by factors that include:

- Stone size

- Location

- Composition

- Individual characteristics (pain threshold, overall health and prior stone experiences)

Signs and symptoms of renal calculi and bladder stones share some similarities, as both conditions involve the urinary tract. Patients with bladder calculi are more likely to have a history of upper tract stones and risk factors for their formation (EAU 2023).

A defining feature of renal calculi is the onset of intense flank pain, known as renal colic, which can radiate to the groin and lower abdomen. The pain can be sudden and intense (see Table 6.4). Haematuria may be observed, either visibly or through microscopic analysis. Urinary symptoms, including increased frequency, urgency, dysuria (painful urination) and a sense of incomplete bladder emptying, can accompany the condition (see Figure 6.3).

Nausea and vomiting may be associated with the severe pain that a person experiences during renal colic. In cases where urinary tract infections develop as a complication, symptoms such as pyrexia and chills may manifest. Large stones or those causing obstruction can lead to hydronephrosis, resulting in enlargement of the kidney due to a build-up of urine and additional discomfort. The specific location of the stone, whether in the ureter or renal pelvis, contributes to the nature of the symptoms, ranging from severe colicky pain to a deep, aching sensation in the flank.

Table 6.4 Renal colic

Cause	The primary cause of renal colic is the obstruction of the urinary tract, usually by kidney stones. The stones can impede the flow of urine from the kidneys to the bladder, leading to a build-up of pressure and causing intense pain.
Location of pain	The pain associated with renal colic is typically located in the flank or lower back, on the side where the affected kidney is located. It can radiate to the groin and lower abdomen.
Character of pain	Renal colic pain is often described as sharp, stabbing, or cramp-like. The intensity of the pain can be excruciating, and it may come in waves.
Onset	The onset of renal colic is usually sudden and can occur without warning. Patients may be relatively asymptomatic until the stone causes an obstruction, which then triggers the pain.
Duration	The duration of renal colic pain can vary. The pain may last for minutes to hours, it may subside temporarily, returning as the stone continues on its journey through the urinary tract.
Associated symptoms	Nausea and vomiting often accompany renal colic and patients may experience restlessness and an inability to find a comfortable position due to the severity of the pain. The patient may sweat profusely.
Pain migration	As the kidney stone moves through the urinary tract, the location of the pain may move. The pain may initially start in the flank and migrate to the lower abdomen or groin as the stone progresses.
Diagnostic importance	The presence of renal colic is a significant clinical indicator of a possible kidney stone. The characteristic pain pattern can help with diagnosis and differentiating kidney stone-related issues from other causes of abdominal or back pain.

Source: Adapted from Jones and Steggall (2019), Fontenelle and Sarti (2019) and Kim and Crook (2021).

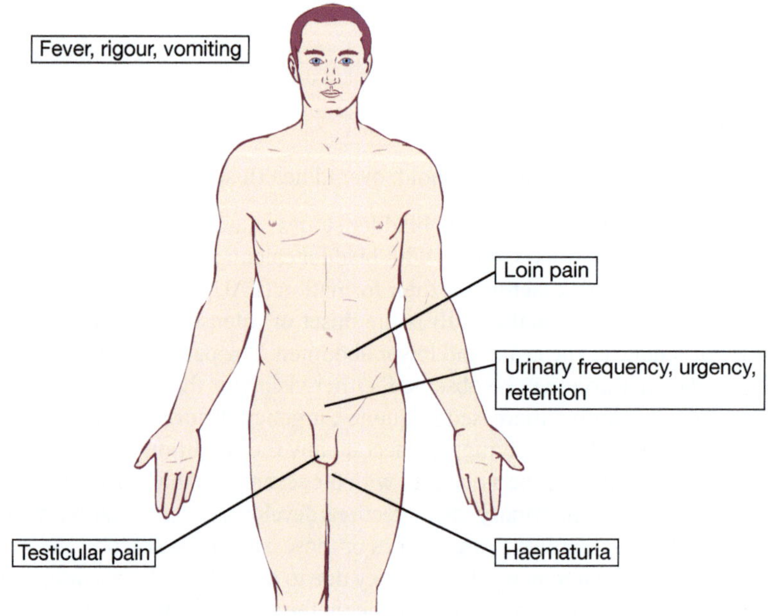

FIGURE 6.3 Signs and symptoms of urinary stones

These are some common clinical features associated with bladder stones:

- Dysuria:

  Individuals with bladder stones may experience discomfort or pain during urination.

- Frequency and urgency:

  Increased frequency of urination and a sense of urgency may be present.

- Haematuria:

  Blood in the urine, either visible or detected through microscopic analysis, can occur due to irritation of the bladder lining.

- Suprapubic pain:

  Pain or discomfort in the lower abdomen, specifically in the suprapubic region (just above the pubic bone), may be felt.

- Difficulty in urination:

  Bladder stones can obstruct the normal flow of urine, leading to difficulty in starting or maintaining a steady stream of urine.

- Incomplete bladder emptying:

  A sensation of incomplete emptying of the bladder, even after urination.

- Lower back pain:

  In some cases, individuals may experience lower back pain.

- Urinary tract infections:

  Bladder stones can contribute to the development of urinary tract infections, leading to symptoms such as fever, chills, and cloudy or foul-smelling urine (maybe due to urinary stasis).

- Visible stones in urine:

  Occasionally, small bladder stones may be passed in the urine, appearing as gritty particles.

Some stones may remain asymptomatic and only come to attention incidentally through imaging studies as they may be so small. Individual responses to kidney stones vary, influenced by factors such as pain tolerance and overall health. Diagnostic tests, including imaging studies and laboratory analyses, are used to help confirm the presence and characteristics of stones.

CLINICAL INVESTIGATIONS AND DIAGNOSIS

Obtaining a patient history, performing a physical examination and requesting laboratory and imaging tests help to make a diagnosis of renal calculi (EAU 2023). Accurate and timely diagnosis of urinary tract calculi is important for prompt clinical management (Kim and Crook 2021).

A careful history may reveal, for example, a number of dietary and lifestyle risk factors that can contribute to a person's risk of stone disease (EAU 2023). Ask the patient about the symptoms experienced, including duration, severity and any exacerbating or relieving factors. The SOCRATES approach can be used (see Table 6.5). Each letter in SOCRATES corresponds to a different aspect of the patient's pain experience.

Table 6.5 Using SOCRATES to assess renal calculi pain

Site (S)	In the context of renal calculi, understanding the site of pain is crucial. Patients may experience pain in the flank (between the ribs and the hip). Pain may radiate towards the lower abdomen and groin as the stone travels through the urinary tract.
Onset (O)	Knowing when the pain began provides insight into the potential causes and progression of renal calculi. Patients might describe a sudden onset of severe pain as the stone moves and causes obstruction or irritation in the urinary tract.
Character (C)	The character of the pain associated with renal calculi is often described as sharp, stabbing or colicky. Renal colic, a hallmark symptom, involves intense spasms of pain as the stone causes intermittent obstruction in the urinary tract.
Radiation (R)	Pain associated with renal calculi can radiate from the flank to other areas, such as the lower abdomen and groin. In women, pain may radiate to the labia or testes in men. The radiation pattern can help in localising the site of the stone within the urinary tract.
Association (A)	Understanding the associated symptoms is crucial for a comprehensive evaluation. Haematuria may indicate irritation in the urinary tract. Patients may also experience nausea and vomiting. Dysuria may occur, especially if the stone is located near the bladder or urethra. Patients might report increased frequency and urgency of urination, signalling irritation caused by the stone. Flank pain is common and may radiate to the lower abdomen or groin.
Time course (T)	The time course of pain is important for assessing the severity and progression of renal calculi. Patients may describe episodes of intense pain during the passage of the stone, with periods of relief in between.
Exacerbating/relieving factors (E)	Identifying factors that worsen or alleviate the pain helps understand the nature of renal calculi. Factors such as increased fluid intake or changes in body position might provide relief. Certain activities, however, or dehydration may exacerbate the pain.
Severity (S)	Renal colic is often described as one of the most severe types of pain. Assessing the intensity of pain using a scale, such as the Numeric Rating Scale or Visual Analogue Scale, allows for a quantitative measure of the pain experience.

Source: Adapted from Jones and Steggall (2019).

Applying the SOCRATES approach to pain related to renal calculi enables a systematic and comprehensive understanding of the patient's symptoms. This information assists in the formulation of a diagnosis, the development of a management plan and the provision of appropriate interventions for pain relief.

In understanding and diagnosing renal calculi, a thorough approach to history taking is imperative. The patient's history becomes a crucial guide in unravelling potential causes and risk factors associated with kidney stones. Thia and Saluja (2021) provide further discussion on history taking when assessing the urinary tract. Discussion with the patient is directed towards the symptoms presented. Ask the patient about the specifics of pain and determine:

- Location

- Intensity

- Radiation

Investigate with the patient any urinary manifestations, such as the presence of blood and any discomfort or pain during urination (dysuria). Establishing the onset and duration of symptoms helps in understanding the temporal progression of the condition. This refers to the development and changes in the condition over time. It involves understanding how the symptoms and characteristics of the condition have evolved since its onset. In the context of renal calculi, assessing the temporal progression would involve determining when the symptoms started, how they have changed and whether there have been any patterns or fluctuations.

A history of recurrent kidney stones should raise red flags. This can then lead to a more in-depth investigation of any underlying metabolic or genetic factors.

To construct a comprehensive understanding, associated symptoms are explored. Nausea, vomiting, fever and chills are examined, as they may indicate complications such as infection or obstruction.

The patient's medical history can reveal any potential contributing factors. Conditions such as gout, urinary tract infections and metabolic disorders may have relevance. Medication history is analysed, with a focus on diuretics or calcium-containing antacids as these may have a possible impact on stone formation (EAU 2023).

Dietary habits, encompassing fluid intake and the consumption of substances such as oxalate or calcium, are probed (EAU 2023). Family history becomes relevant, considering the genetic/hereditary aspect of kidney stones (Fontenelle and Sarti 2019).

Occupational and environmental exposure should not be overlooked. Certain professions and environmental conditions may heighten the risk of kidney stone formation. A holistic approach extends to the patient's lifestyle, encompassing physical activity, as sedentary behaviour is acknowledged as a potential risk factor.

This thorough investigation forms the basis for a detailed understanding, assisting with the formulation of a differential diagnosis. Confirmatory steps often involve physical examination, imaging studies and laboratory tests. The patient's own story, their narrative linked with these details, guides accurate diagnosis and effective management.

PHYSICAL EXAMINATION

A full and thorough abdominal examination should be performed to help exclude differential diagnoses, for example, ruptured aortic aneurysm, appendicitis, diverticulitis and peritonitis. The person may be restless and unable to lie still (this observation helps differentiate renal colic from peritonitis), (Thia and Saluja 2021).

Observe for signs which may indicate a complication, such as hesitancy of micturition or an intermittent urinary stream, as this may suggest urinary tract obstruction. If there is pyrexia and sweating, this can indicate a coexisting urinary tract infection.

With regards to bladder stones, the EAU (2023) suggests the following should be performed where possible prior to (or at the time of) bladder stone treatment, physical examination of external genitalia and peripheral nervous system (including digital rectal examination, peri-anal tone and sensation in men).

Maintaining dignity and ensuring the comfort of a patient during a physical examination is essential for fostering trust, cooperation and a positive healthcare experience. Every patient has the right to be treated with respect and dignity. During the examination, local policy and procedure must be adhered to with regards to infection prevention and control. The patient should be offered the presence of a chaperone (General Medical Council 2013). When possible, consider the patient's gender preferences when providing this. Adhere to local policy and procedure and ensure that informed consent is gained.

Clearly communicate the purpose of each step in the examination, informing the patient about what to expect. This helps reduce anxiety and allows the patient to prepare mentally. Always obtain informed consent before proceeding with any aspect of the examination. Explain the nature of the examination and obtain the patient's agreement to proceed. Ensure that the examination is undertaken in a private area and use curtains or doors to create a private space. Give the patient the opportunity to undress and dress in private. Use drapes or gowns to cover areas of the body not being examined, exposing only the necessary parts. This helps maintain the patient's modesty. This helps the patient feel more in control and less vulnerable.

Involve the patient in the process by explaining the purpose of the examination and encouraging them to ask questions or express concerns. Be attentive to the patient's emotional state and offer support throughout the examination. This may include providing reassurance, acknowledging any discomfort and addressing concerns.

Adhering to the principles discussed, a respectful and patient-centred environment can be created. This approach ensures that individuals maintain their dignity and feel comfortable throughout the physical examination process.

The abdominal examination includes gentle palpation to identify tenderness, guarding or palpable masses, focusing on the kidneys' area. The costovertebral angle tenderness test, involving percussion in the area formed by the spine and the 12th rib, is key for detecting kidney-related discomfort (Shaw 2016).

The genitourinary examination includes inspecting and palpating the genitalia to identify signs of discomfort or tenderness that may indicate referred pain from renal calculi. A neurological examination assesses the patient's overall neurological status, considering that severe pain from renal colic can lead to restlessness, agitation and autonomic responses such as sweating.

Vital signs, such as blood pressure and heart rate, are monitored to evaluate the patient's general condition, recognising that pain associated with renal calculi may cause fluctuations in these parameters. Skin examination for signs of dehydration, for example, dryness or poor turgor, is relevant since dehydration is a risk factor for kidney stone formation. Skin turgor can be assessed by gently pinching two folds of skin which should return to their normal position; during dehydration, the skin may take longer to return to its normal state. The skin turgor test is not reliable for older adults as skin turgor takes longer to return to normal even in a healthy patient (McErlean 2021).

While a physical examination provides valuable clinical information, imaging studies are often required to confirm the presence of renal calculi, assess their characteristics and identify potential complications. The physical examination, when combined with a comprehensive patient history and additional diagnostic measures, plays a crucial role in guiding the diagnosis and developing an appropriate treatment plan for individuals with suspected renal calculi.

CLINICAL INVESTIGATIONS

Investigations and tests may be required based on the initial findings (the history and examination).

Urine dipstick testing may be used in the diagnosis of renal calculi, or kidney stones. This can provide support for the diagnosis and help to exclude infection. It can detect the presence of blood in the urine, which can also support the diagnosis of renal or ureteric colic. It should be noted, however, that the absence of haematuria does not exclude a diagnosis but should prompt consideration for other causes of pain (EAU 2023). Urine dipstick testing can be a useful tool but should be used in conjunction with other diagnostic methods, for example, imaging studies.

A summary of the clinical investigations and tests that may be used in the diagnosis of renal calculi is outlined in Table 6.6. Pain relief, or any other emergency measures, should not be delayed by undertaking the imaging assessment (EAU 2023).

Table 6.6   An overview of tests and investigations used in making a diagnosis of renal calculi

Test/investigation	Discussion
Imaging studies	The most appropriate imaging modality is decided by the clinical situation, which differs depending on whether a ureteral or a renal stone is suspected. Urgent imaging is required within 24 hours.
	Ultrasound: Ultrasound is a non-invasive imaging modality that can visualise the kidneys and urinary tract. It is particularly useful for detecting larger stones and assessing the overall anatomy of the kidneys. This is the primary diagnostic tool.
	Non-contrast computed tomography (CT) scans: These scans are highly sensitive and specific for detecting renal calculi. CT scans can accurately identify the location, size and number of stones. This imaging technique is often considered the gold standard for diagnosing kidney stones due to its high resolution.
Laboratory tests	Serum creatinine: Elevated levels of serum creatinine may indicate impaired kidney function. While not specific to kidney stones, assessing kidney function is essential, especially if there are concerns about obstruction or infection.
	Stone analysis: If a stone is passed or removed, analysing its composition can provide insights into the type of stones present (e.g. calcium oxalate, struvite, uric acid). This information may guide preventive measures.
	Biochemical work-up is similar for all stone patients. Serum blood tests such as: • Creatinine • Uric acid • Calcium • Sodium • Potassium • Blood cell count • C-reactive protein However, if no intervention is planned, examination of sodium, potassium, C-reactive protein and blood coagulation time can be omitted. Only patients at high risk for stone recurrence should undergo a more specific analytical programme.
Specialised investigations	24-hour urine collection: This test assesses the concentration of various substances in the urine over a 24-hour period. It may help identify metabolic abnormalities contributing to stone formation, such as high levels of calcium, oxalate or uric acid.
Cystoscopy	In certain cases, a cystoscopy may be performed to directly visualise the inside of the bladder and urethra (see Figure 6.4). This is more relevant if lower urinary tract symptoms or recurrent urinary tract infections are present.

Source: Adapted from EAU (2023).

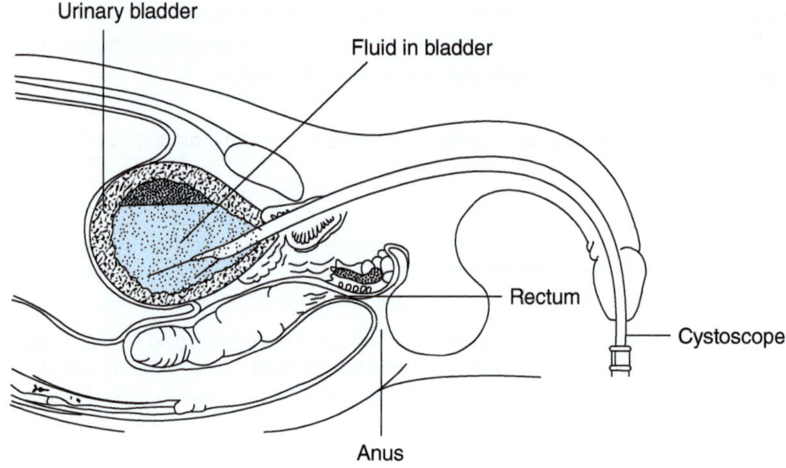

FIGURE 6.4 Cystoscopy

MANAGEMENT

The overall goals of management for those with renal calculi are to provide the patient with pain relief and to relieve obstruction. The treatment of renal calculi is based on a number of parameters and is individualised to meet the needs of each patient. Parameters such as the size, number, location and constitution of the stones are important for deciding the treatment. Furthermore, the morphology, shape, form, structure, volume, mobility, configuration and hardness of the stone should be considered. For each patient, it is important to examine and understand the structure (anatomy) of the pelvis and calyces in the kidneys. The pelvic-calyceal system is a network of tubes in the kidney where urine collects before moving to the bladder; assessing this system helps in understanding any potential issues or conditions affecting the kidneys (EAU 2023).

Non-steroidal anti-inflammatory drugs, usually in the form of diclofenac intramuscular or per rectum, should be offered as first-line for the relief of the severe pain of renal colic. Non-steroidal anti-inflammatory drugs are more effective than opioids and have less tendency to cause nausea; if they are contra-indicated, offer intravenous paracetamol. If both are contra-indicated or are ineffective, opiates are considered (pethidine is avoided as this is more likely than other options to induce vomiting).

Antiemetics and rehydration therapy are provided, if needed; there is no place for antispasmodics.

Further management will depend on the size of the stone, the likelihood of spontaneous passage, whether the stone is renal or ureteric, the severity of symptoms, the patient's age and any relevant contra-indications or comorbidities.

Watchful waiting should be considered for asymptomatic renal stones in adults, children and young people if the stone is:

- Less than 5 mm.

- Larger than 5 mm and the person consents to watchful waiting after an informed discussion of the possible risks and benefits.

Medical expulsive therapy may be used to facilitate the passage of the stone. It is considered in cases where the stone is located distally. The optimal size is greater than 5 mm but less than 10 mm in diameter. Alpha-blockers (e.g. tamsulosin) are given (EAU 2023).

Table 6.7 offers an overview of the management of renal calculi.

The management of a person with renal calculi is individualised and the choice of intervention depends on the specific characteristics of the stones and the patient's overall health. A multidisciplinary collaborative approach is advocated, which involves urologists, nephrologists and dietitians, essential in the provision of holistic patient care and preventing stone recurrence.

Table 6.7 Management of renal calculi

Hydration	Increased fluid intake: Adequate hydration is crucial for flushing out the urinary system and promoting the natural passage of stones. Patients are advised to drink plenty of water to maintain a high urine output.
Observation and monitoring	Imaging studies: Regular monitoring through imaging studies helps track the position and movement of stones. Adjustments to the treatment plan can be made based on the patient's response.
Dietary modifications	Low oxalate diet: For certain types of stones, a low-oxalate diet may be recommended. Dietary calcium: Adequate dietary calcium intake may help prevent the formation of certain types of stones.
Extracorporeal shock wave lithotripsy	Shock wave therapy: Extracorporeal shock wave lithotripsy uses shock waves to break larger stones into smaller fragments, making them easier to pass. It is a non-invasive procedure, often suitable for smaller stones.
Ureteroscopy	Endoscopic stone removal: Ureteroscopy involves passing a thin tube through the urethra and bladder to reach the stone in the ureter or kidney. The stone can then be removed or fragmented using laser energy.
Percutaneous nephrolithotomy	Minimally invasive surgery: Percutaneous nephrolithotomy is a surgical procedure that involves making a small incision in the back to access and remove larger stones from the kidney.
Preventive measures	Lifestyle modifications: Addressing modifiable risk factors, for example, inadequate fluid intake, dietary habits and sedentary behaviour. Medication for underlying conditions: Treating conditions such as hyperparathyroidism or gout, if contributing to stone formation.
Follow-up	Regular monitoring: Patients with a history of kidney stones may undergo regular follow-up, including imaging studies and metabolic evaluations, to detect and address new stones or underlying metabolic abnormalities.
Patient education	Lifestyle guidance: Advising patients about dietary choices, fluid intake and lifestyle changes to reduce the risk of recurrent stone formation.

HEALTH TEACHING

Health education for patients with renal calculi is key to empowering them with knowledge about their condition, promoting preventive measures and enabling better management. Lifestyle modifications, including regular physical activity, stress management and smoking cessation, are encouraged in order to promote overall kidney health.

Follow-up care and regular check-ups with healthcare providers are stressed, ensuring ongoing monitoring and timely intervention. Providing patient-friendly resources, such as written materials and online tools, enhances comprehension and supports continuous patient education.

Ultimately, the goal of health education for patients with renal calculi is to empower individuals, encourage positive lifestyle changes, and foster a proactive approach to managing and preventing kidney stones.

CONCLUSION

This chapter has provided essential knowledge to understand, assess and care for individuals affected by renal calculi; the pathophysiology, risk factors, clinical manifestations and diagnostic approaches associated with renal calculi have been outlined.

The importance of obtaining a detailed patient history, undertaking thorough physical examination and appropriate use of clinical investigations in arriving at an accurate diagnosis has been emphasised. The integration of the SOCRATES approach to pain and the consideration of temporal progression provide a structured framework for effective history taking in the context of renal calculi.

The emphasis on maintaining the dignity and comfort of patients during physical examinations is a principle that should be upheld. Acknowledging the sensitivity of this condition, the chapter has highlighted the importance of effective communication, privacy and patient empowerment in delivering compassionate care.

A range of management strategies were discussed. Understanding these diverse approaches can help to contribute to comprehensive patient care plans and enhance collaboration within interdisciplinary healthcare teams.

GLOSSARY OF TERMS

Calcium oxalate: A common type of kidney stone composed of calcium and oxalate crystals.

Calyx (plural: calyces): Cup-like structures in the kidney where urine collects before flowing into the renal pelvis.

Cystoscopy: A medical procedure involving using a thin tube with a camera to examine the urethra and bladder.

Dysuria: Pain or discomfort during urination.

Extracorporeal shock wave lithotripsy: Non-invasive procedure using shock waves to break kidney stones into smaller pieces for easier passage.

Flank pain: Pain often associated with kidney stones.

Haematuria: Presence of blood in the urine.

Lithotripsy: Medical procedure to break down kidney stones using sound waves or shock waves.

Medical expulsive therapy: Treatment to facilitate the natural passage of kidney stones, often involving medications to relax the ureter.

Nephrolithiasis: Term for the presence of kidney stones.

Oxalate: A substance found in many foods that can contribute to the formation of certain types of kidney stones.

Percutaneous nephrolithotomy: A surgical procedure to remove kidney stones through a small incision in the back.

Renal colic: Severe pain caused by the obstruction of urine flow due to a kidney stone.

SOCRATES: An acronym used in medical history taking to explore the:

Site
Onset
Character
Radiation
Association
Time course
Exacerbating/relieving factors
Severity of symptoms

Renal pelvis: The central collecting region in the kidney where urine is formed before flowing into the ureter.

Thiazide diuretics: Medications that promote diuresis and are sometimes used to prevent certain types of kidney stones.

Ureteroscopy: Endoscopic procedure using a thin tube to visualise and treat kidney stones.

Urinalysis: Examination of urine to detect abnormalities, including the presence of blood or crystals.

Urinary tract infection: Infection involving the urinary system, which may be associated with kidney stones.

Urolithiasis: Formation of stones in the urinary system, including the kidneys.

MULTIPLE CHOICE QUESTIONS

1. What is the primary component of calcium oxalate stones, a common type of renal calculi?
   a) Uric acid
   b) Calcium
   c) Oxalate
   d) Phosphate

2. Which imaging technique is considered the gold standard for diagnosing renal calculi?
   a) Ultrasound
   b) X-ray
   c) Non-contrast CT scan
   d) Intravenous urography

3. What does the term 'renal colic' refer to in the context of renal calculi?
   a) Inflammation of the kidneys
   b) Pain associated with kidney stones
   c) Formation of crystals in the urine
   d) Urinary tract infection

4. Which medication is commonly used in medical expulsive therapy to aid the passage of kidney stones?
   a) Prednisone
   b) Tamsulosin
   c) Ibuprofen
   d) Nifedipine

5. Which term refers to tenderness at the costovertebral angle and is associated with kidney stone pain?
   a) Epigastric tenderness
   b) Flank tenderness
   c) Suprapubic tenderness
   d) Inguinal tenderness

6. What surgical procedure involves making a small incision in the back to access and remove kidney stones?
   a) Ureteroscopy
   b) Extracorporeal shock wave lithotripsy
   c) Percutaneous nephrolithotomy
   d) Cystoscopy

7. What does the acronym SOCRATES stand for in the context of history taking for renal calculi?
   a) Symptoms, Onset, Characteristics, Radiation, Association, Time, Evaluation, Severity
   b) Site, Onset, Characteristics, Radiation, Association, Time course, Exacerbating/relieving factors, Severity
   c) Systematic, Onset, Characteristics, Radiation, Association, Timing, Evaluation, Severity
   d) Site, Orientation, Characteristics, Radiation, Association, Time course, Evaluation, Severity

8. Which medication is commonly used to manage pain associated with renal calculi?
   a) Antibiotics
   b) Antihypertensives
   c) Non-steroidal anti-inflammatory drugs
   d) Antipyretics

9. What is the term for the cup-like structures in the kidney where urine collects before flowing into the renal pelvis?
   a) Renal pelvis
   b) Calyx (calyces)
   c) Ureter
   d) Nephron

10. In which part of the urinary system do stones typically cause obstruction leading to renal colic?
    a) Urethra
    b) Bladder
    c) Ureter
    d) Renal pelvis

REFERENCES

Alelign, T. and Petros, B. (2019). Kidney stone disease: an update on current concepts. *Advances in Urology.* doi: 10.1155/2018/3068365.

Dawson, C.H. and Thomas, C.R.V. (2012). Kidney stone disease: pathophysiology, investigation and medical treatment. *Clinical Medical Journal Royal College of Physicians* (London). doi: 10.7861/clinmedicine.

European Association of Urology (2023). EAU guidelines on urolithiasis. https://d56bochluxqnz.cloudfront.net/documents/full-guideline/EAU-Guidelines-on-Urolithiasis-2023.pdf (accessed November 2023).

Fontenelle, L.F. and Sarti, T.D. (2019). Kidney stones: treatment and prevention. *American Family Physician* 99 (8): 490–496.

General Medical Council (2013). Intimate examinations and chaperones. https://www.gmc-uk.org/-/media/documents/maintaining-boundaries-intimate-examinations-and-chaperones_pdf-58835231.pdf (accessed November 2023).

Hagler, D., Harding, M.M., Kwong, J. et al. (2023). *Lewis's Medical Surgical Nursing*, 12e. St Louis: Elsevier.

Jones, K. and Steggall, M. (2019). Nursing patients with urinary disorders (Chapter 9). In: *Alexanders Nursing Practice,* 5e (ed. I. Peate). London: Elsevier.

Kim, E.J. and Crook, M.A. (2021). Urolithiasis: don't forget the rarities. *Annals of Clinical Biochemistry* 58 (5): 392–394. doi: 10.1177/00045632211018711 journals.sagepub.com/home/acb.

Lopez, M. and Hoppe, B. (2008). History, epidemiology and regional diversities of urolithiasis. *Pediatric Nephrology* 25 (1): 49–59. doi: 10.1007/s00467-008-0960-5.

McErlean, L. (2021). Fluid balance (Chapter 32). In: *The Nursing Associate's Handbook of Clinical Skills* (ed. I. Peate). Oxford: Wiley.

Nagalingam, K. (2021). The renal system and associated disorders (Chapter 11). In: *Fundamentals of Pathophysiology Anatomy and Physiology.* 4e (ed. I. Peate). Oxford: Wiley.

O'Callaghan, C. (2016). *The Renal System at a Glance,* 4e. Oxford: Wiley.

Public Health England and Industrial Injuries Advisory Council (2018). Occupational risks for urolithiasis: IIAC information note. https://www.gov.uk/government/publications/occupational-risks-for-urolithiasis-iiac-information-note/occupational-risks-for-urolithiasis-iiac-information-note (accessed November 2023).

Shang, W., Li, L., Ren, Y. et al. (2017). History of kidney stones and risk of chronic kidney disease: a meta-analysis. *Peer J Publishing* 5: e2907. doi: 10.7717/peerj.2907.

Shaw, M. (2016). Urolithiasis (Chapter 6). In: *Urology at a Glance* (eds. H. Hashim and P. Dasgupta). Oxford: Wiley.

Swales, J. (2022). The person with a urinary disorder (Chapter 29). In: *Nursing Practice,* 3e. (eds. I. Peate and A. Mitchell). Oxford: Wiley.

Thia, I. and Saluja, M. (2021). An update on the management of renal colic. *Australian Journal of General Practice* 50 (7): 445–449.

Yaqoob, M.M. and Ashman, N. (2021). Kidney and urinary tract disease (Chapter 36). In: *Kumar and Clark's Clinical Medicine,* 10e. (eds. A. Feather, D. Randall and M. Waterhouse). London: Elsevier.

Urological cancer is an umbrella term that is used to refer to cancers that affect the organs of the urinary system and the male reproductive system. They may also be known as urothelial tumours or genitourinary cancers. These cancers can occur in various parts of the urinary tract and reproductive organs. Table 7.1 outlines the main types of urological cancers (excluding bladder cancer).

Table 7.1 Main types of urological cancer (excluding bladder cancer)

Type of urological cancer	Discussion
Prostate cancer	Prostate cancer usually presents with lower urinary tract symptoms, including nocturia, urinary frequency and hesitancy. Haematuria can occur, as can erectile dysfunction. Some prostate cancers present with disseminated disease, typically metastases to bone.
	The lower urinary symptoms overlap with those of benign prostatic hyperplasia – and the two conditions can co-exist. Digital rectal examination can help to differentiate the two, with hardness of the prostate or individual nodules being features suggestive of cancer.
	Prostate-specific antigen testing is generally available in primary care, with age-specific raised values suggestive of cancer.
	A definitive diagnosis requires biopsy, often guided by imaging. This is performed in a secondary care setting.
Renal cancer	Renal cancer symptoms include haematuria, loin pain, urinary tract infections or a mass in the flank.
	The symptoms overlap with other urological cancers, particularly bladder cancer.
	Most renal cancers are visible on ultrasound of the kidneys – a test that is available on referral to primary care.
	Definitive diagnosis of renal cancer requires histology, performed in secondary care.
Testicular cancer	Testicular cancer usually presents as a change in the shape or texture of the testis. This may be painful.
	It can present as disseminated disease, particularly with lymph node spread.
	Testicular cancer can be seen on ultrasound of the testis, an investigation available in primary care settings.
Penile cancer	Penile cancer is usually seen as a raised lesion. Penile cancers are rare.
	It can be difficult to differentiate penile cancer from the common lesions seen with some sexually transmitted infections.
	It is often possible to diagnose a typical penile cancer visually, but confirmation of the diagnosis is generally made by excision biopsy in secondary care.

Source: Adapted from Leach (2023).

There are a number of risk factors associated with urological cancers, so early detection and appropriate treatment are crucial for better outcomes. Symptoms vary depending on the type and stage of the cancer but may include changes in urinary habits, blood in the urine, pain or lumps in the affected organs.

Bladder cancer is a multifaceted condition. Its manifestations range from subtle disruptions in urinary patterns to more worrying signs, necessitating a nuanced understanding for prompt identification and intervention. This chapter equips the reader with insights into the complexities of bladder cancer, from understanding its pathophysiology and risk factors to developing the skills needed to offer people effective care.

Bladder cancer brings with it various presentations and treatment modalities. Key to treatment and care provision is the importance of a holistic approach that must consider not only the physical aspects of the condition but also the emotional and psychological dimensions that may be experienced by individuals facing a bladder cancer diagnosis.

Sometimes the terms' cancer, tumour and neoplasm are used interchangeably. These terms are related but they have distinct meanings (see Table 7.2).

This chapter will focus on bladder cancer. Bladder cancer can occur at any age but is more common in older adults. Bladder cancer is a type of cancer that begins in the cells lining the urinary bladder, a hollow, distensible organ that is located in the pelvis, posterior to the symphysis pubis. The bladder's primary function is to store urine that has been produced by the kidneys until it is ready to be excreted from the body via the urethra. The bladder has a folded internal lining, known as rugae, which allows it to store up to 400–600 mL of urine in healthy adults.

Table 7.2   Cancer, tumour and neoplasm

Term	Definition	Characteristics	Summary
Neoplasm	Neoplasm is a broad term that refers to an abnormal and uncontrolled growth of cells. It can be benign (not cancerous) or malignant (cancerous).	Neoplasms can develop in various tissues and organs of the body. They are characterised by the abnormal proliferation of cells that may form a mass or lump.	General term for abnormal cell growth, can be benign or malignant.
Cancer	Cancer specifically refers to a malignant neoplasm, indicating a group of diseases characterised by the uncontrolled growth and spread of abnormal cells.	Cancerous cells can invade surrounding tissues and in some cases, metastasise to other parts of the body.	A group of diseases characterised by uncontrolled cell growth and potential invasion and metastasis.
Tumour	Tumour is a more general term and can refer to both benign and malignant neoplasms. It is often used interchangeably with 'mass' or 'lump'.	Tumours can be classified as either benign or malignant. Benign tumours are usually localised, slow-growing and non-invasive. Malignant tumours, on the other hand, have the potential to invade nearby tissues and spread to other parts of the body.	General term for abnormal growths, can be benign or malignant.

PATHOPHYSIOLOGICAL CHANGES ASSOCIATED WITH BLADDER CANCER

Bladder cancer is caused by changes in the cells of the bladder, which can be linked with exposure to certain chemicals. There are changes in how bladder cells function, particularly how they grow and divide into new cells (National Cancer Institute 2023).

Bladder cancer develops through a series of complex pathophysiological changes within the bladder, progressing through initiation, promotion and progression stages. The intricate journey of abnormal cell proliferation begins with initiation, which is often triggered by a combination of genetic and environmental factors. Particularly, exposure to tobacco smoke plays a significant role, as carcinogens that are present in tobacco interact with the bladder lining, initiating genetic mutations in crucial genes, for example, TP53 and HRAS. These genes play important roles in regulating cell growth, preventing the development of tumours and maintaining the overall stability of the cellular environment.

Understanding the roles of these crucial genes can provide insights into the molecular mechanisms that underlie bladder cancer. This can guide targeted therapeutic approaches that are aimed at restoring normal cellular regulation. It is important to note that bladder cancer is a heterogeneous disease (a disease that exhibits diversity or variability) and alterations in various genes can contribute to its development and progression.

The subsequent promotion stage associated with bladder cancer sees the uncontrolled growth of mutated cells. This leads to the development of pre-malignant lesions and an abnormal increase in the number of cells in the bladder (atypical hyperplasia). Prolonged exposure to carcinogens and chronic inflammation, which is often associated with conditions such as recurrent urinary tract infections, contributes to the progression of this phase, exacerbates DNA damage and promotes cellular proliferation. This damage can lead to mutations that cause the cells to grow and divide uncontrollably, forming a tumour.

The culmination of this progression is the invasion and potential metastasis of cancerous cells, particularly when the malignancy infiltrates the muscular layer, the deeper layers of the bladder wall. This invasive potential will increase the risk of metastasis to the nearby lymph nodes as well as distant organs via the lymphatic system. Simultaneously, the tumour will stimulate angiogenesis, the process of generating new blood vessels, to support its continued growth. Additionally, the process of epithelial–mesenchymal transition (EMT) transforms cells, enhancing their ability to migrate and invade surrounding tissues. As the cancer progresses, it can cause a range of symptoms.

Understanding these intricate and complex pathophysiological changes is essential for tailoring targeted therapies, developing effective early detection strategies and providing care. Regular screening, particularly for those individuals at high risk, are important for timely identification and intervention in bladder cancer.

Table 7.3 provides an overview of the pathogenesis of bladder cancer.

Bladder cancer is usually transitional cell (urothelial) carcinoma. Cancer originates from the transitional cells of the mucosal urothelium, the specialised epithelial lining that forms the innermost layer of the bladder. This epithelium is primarily composed of transitional cells, also known as urothelial or uroepithelial cells. This can appear as either a non-invasive, papillary tumour on the mucosal surface or as a solid, non-papillary tumour that infiltrates the bladder wall, with an increased likelihood of metastasis. Non-papillary tumours in the bladder arise from a condition called in situ dysplasia. In situ dysplasia refers to abnormal changes in the cells of the mucosal urothelium without invasion into surrounding tissues. These abnormal cells can give rise to solid, non-papillary tumours that may have a higher likelihood of invading the bladder wall and potentially spreading to other parts of the body.

Table 7.3   Pathogenesis of bladder cancer, an overview

Stage	Pathogenesis
Initiation	Genetic factors:
	Genetic alterations play a significant role in the initiation of bladder cancer. Exposure to carcinogens, such as those found in tobacco smoke and certain occupational exposures, can lead to mutations in critical genes. The tumour suppressor gene *TP53* and oncogenes such as HRAS are commonly implicated in bladder cancer.
	Environmental exposures:
	Tobacco smoke is a well-established risk factor for bladder cancer. Carcinogens present in tobacco can be excreted through urine, exposing the bladder lining to harmful substances and increasing the risk of genetic mutations.
Promotion	Chronic inflammation:
	Conditions leading to chronic inflammation of the bladder, such as recurrent urinary tract infections or long-term catheter use, can promote bladder cancer. Inflammation contributes to DNA damage, cell proliferation and the formation of pre-malignant lesions.
	Chemical exposures:
	Occupational exposures to certain chemicals, such as aromatic amines and polycyclic aromatic hydrocarbons, have been linked to an increased risk of bladder cancer. These chemicals can enter the body and come into contact with the bladder lining.
Progression	Cellular changes:
	As the disease progresses, genetic mutations accumulate, leading to the transformation of normal urothelial cells into cancerous cells. Mutations in genes involved in cell cycle regulation, apoptosis and DNA repair contribute to uncontrolled cell growth.
	Invasion and metastasis:
	Bladder cancer can advance to invade the deeper layers of the bladder wall, including the muscle layer. Muscle-invasion is a critical step associated with a higher risk of metastasis to lymph nodes and distant organs.
	Angiogenesis:
	Tumour cells stimulate the formation of new blood vessels (angiogenesis) to ensure a blood supply for their growth and progression. This process facilitates the delivery of oxygen and nutrients to the tumour.
	Epithelial–mesenchymal transition (EMT):
	EMT is a process in which cancer cells acquire characteristics that enhance their migratory and invasive capabilities. This contributes to the spread of cancerous cells beyond the bladder.

EPIDEMIOLOGY

In-depth bladder cancer statistics are provided by Cancer Research UK (CRUK). See more in-depth bladder cancer incidence statistics at Bladder cancer incidence statistics | CRUK.

Between 2016 and 2018, there were approximately 10 300 new bladder cancer cases in the UK every year. This equates to 28 cases every day. In the UK, bladder cancer is the 11th most common cancer. It accounted for 3% of all new cancer cases between 2016 and 2018. In women, this is the 16th most common cancer, with around 2800 new cases every year and in men, it is the eighth most common cancer, with about 7500 new cases every year. Incidence rates for bladder cancer are highest in people aged 85–89 years.

Between 2016 and 2018, rates in women have decreased by around two-fifths (41%) and rates in men have decreased by almost half (47%) (2016–2018). In England, around 980 cases of bladder cancer each year are linked with deprivation (around 370 in women and around 610 in men).

In England (2013–2017), incidence rates for bladder cancer are lower in the Asian and Black ethnic groups, compared with the White ethnic group.

RISK FACTORS

Bladder cancer risk factors are diverse and can involve a combination of genetic, environmental and lifestyle factors. Table 7.4 provides a discussion of key risk factors associated with the development of bladder cancer.

CLINICAL PRESENTATION

Most patients present with unexplained haematuria (visible or microscopic). Blood in the urine is the most common symptom of bladder cancer. Around 80 out of 100 people with bladder cancer (around 80%) have some blood in their urine (CRUK 2022a). Some patients present with anaemia and haematuria can be detected as part of other investigations - a coincidental finding.

Table 7.4 Key risk factors for bladder cancer

Risk factor	Discussion
Tobacco use	Cigarette smoking is the most significant risk factor for bladder cancer. Chemicals in tobacco smoke are excreted in the urine and can directly affect the bladder lining. Around half of all bladder cancers are caused by smoking.
Occupational exposures	Certain types of jobs can carry higher risk than others, depending on the exposures people have in their jobs. Exposure to certain chemicals and substances in the workplace, such as aromatic amines and industrial chemicals, is linked to an increased risk of bladder cancer.
Age and gender	Bladder cancer risk increases with age, and it is more common in men than women.
Race and ethnicity	Caucasians are more likely to develop bladder cancer than other racial or ethnic groups.
Previous cancer treatment	Individuals who have undergone certain cancer treatments, such as radiation therapy or chemotherapy with cyclophosphamide, may have an elevated risk of developing bladder cancer.
Chronic bladder inflammation and infections	Long-term inflammation or infections of the urinary bladder, such as those related to urinary stones or recurrent urinary tract infections, may increase the risk.
Genetic factors	Individuals with a family history of bladder cancer may have a higher risk, suggesting a potential genetic predisposition.
Personal history of bladder cancer	Individuals who have had bladder cancer in the past have a higher risk of developing new bladder cancers.
Chronic bladder irritation	Prolonged exposure to substances that irritate the bladder, such as certain medications or catheters, may contribute to an increased risk.

Source: Adapted from CRUK (2023); Centers for Disease Control and Prevention (2023).

Irritative voiding symptoms can include passing urine very often (frequency), passing urine very suddenly (urgency), pain or a burning sensation when passing urine (dysuria) are also common at presentation as well as pyuria (pus in the urine). Pelvic pain, bone pain and back pain occurs with advanced cancer, a pelvic mass may be palpable. Other symptoms include weight loss for no reason, feeling tired and unwell. See Table 7.5 for a summary of the clinical features.

The clinical presentation of bladder cancer varies depending on factors such as the stage and grade of the tumour. Early-stage bladder cancer may present with different symptoms compared to more advanced stages.

The symptoms discussed are not exclusive to bladder cancer and can be caused by various other conditions, including urinary tract infections and benign prostatic hyperplasia. However, the presence of persistent symptoms, especially haematuria, should prompt further investigation. The clinical presentation may differ between non-muscle-invasive bladder cancer (confined to the inner layers of the bladder) and muscle-invasive bladder cancer (spread to the muscle layer or beyond). Early detection is key to improving outcomes and facilitating effective treatment.

CLINICAL INVESTIGATIONS AND DIAGNOSIS

Bladder cancer is usually diagnosed by going through the patient's health history, performing a physical examination and observing for signs of the disease. Emergency admission to the hospital is also a common way for bladder cancer to present and is often associated with a poor prognosis (National Institute for Health and Care Excellence [NICE] 2021).

Cystoscopy is the standard investigation for suspected bladder cancer and for treatment follow-up. If abnormalities are found during cystoscopy, white light-guided

Table 7.5 A summary of the clinical features associated with bladder cancer

Clinical feature	Discussion
Haematuria	• Visible haematuria, where blood is noticeable in the urine and may give it a pink or red colour, is a hallmark symptom of bladder cancer.
	• Microscopic haematuria, which is not visible to the naked eye but can be detected through laboratory tests, is also common.
Urinary symptoms	• Increased frequency of urination.
	• Urgency to urinate, often with a feeling of being unable to hold it.
	• Pain or a burning sensation during urination (dysuria).
Pelvic pain or discomfort	• Pain in the lower abdomen or pelvic region may be experienced, especially as the tumour grows or if there is an invasion into nearby structures.
Back or flank pain	• Pain in the lower back or flank area may occur if the cancer has invaded the surrounding tissues or structures.
Changes in urinary habits	• Changes in urinary habits, such as difficulty starting or maintaining a steady stream of urine, may be indicative of bladder cancer.
Incomplete emptying of the bladder	• The patient may report a sensation of incomplete emptying of the bladder, even after urination.
Systemic symptoms (advanced cases)	• In more advanced stages, bladder cancer can lead to systemic symptoms such as weight loss, fatigue and loss of appetite.

TURBT is recommended. If muscle-invasive bladder cancer is suspected at cystoscopy, computed tomography (CT) or magnetic resonance imaging (MRI) staging should be considered before TURBT.

HISTORY TAKING

When taking a patient history from a patient who may have bladder cancer, it is crucial to gather comprehensive information that can aid in the diagnosis and subsequent management of the condition. It is essential to establish rapport with the patient. Begin the conversation by introducing yourself, explaining your role and creating a comfortable environment.

Inquire about the reason for their visit, prompting them to describe their chief complaint. This must be done with sensitivity and with compassion. Encourage an open discussion about symptoms related to bladder cancer, focusing on aspects such as changes in urinary habits, pelvic pain and other concerning signs.

For those patients who report haematuria, ask about its onset, frequency and visibility. Explore factors that may exacerbate or alleviate the haematuria. Investigate changes in urinary habits, such as frequency, urgency, dysuria and feelings of incomplete bladder emptying. Pay close attention to any pain or discomfort the person describes, identify its location, intensity and factors influencing the symptoms, particularly in the lower abdomen, pelvic region or the back. Inquire about systemic symptoms such as weight loss, fatigue or loss of appetite, which could indicate more advanced stages of bladder cancer.

As the consultation progresses, focus on potential risk factors, addressing areas such as tobacco use, occupational exposures, family history of bladder cancer and any history of radiation or chemotherapy, use a systematic approach. If applicable, discuss the patient's occupation, identifying any potential exposure to industrial or chemical substances associated with an increased risk of bladder cancer. Seek information about the person's overall health, existing medical conditions, history of cancer and any past surgery or treatments. Document their current medication regimen.

Allow the patient to express their concerns and expectations regarding their health and current symptoms. Consider the patient's psychosocial history, including any stressors or emotional factors that might impact their health. Summarise key information obtained during the interview, ensuring clarity on the importance of the details discussed. Discuss the next steps, including any necessary examinations or tests that may be required. Throughout the process, demonstrate empathy, active listening and sensitivity to the patient's concerns. A patient-centred approach and effective communication are essential for building trust and facilitating optimal care.

PHYSICAL EXAMINATION

The physical examination for a patient suspected of having bladder cancer is an essential component of the diagnostic process. It complements other diagnostic tests and helps assess signs and symptoms that may indicate bladder abnormalities. Physical examination is not only about gathering clinical information. It is also an opportunity to build a therapeutic relationship with the patient (Senior 2021).

The examination begins with an introduction, explaining the role of the person undertaking the examination and the creation of a comfortable and respectful environment, giving full attention to the patient throughout the examination. A chaperone is offered and local policies and procedures regarding infection prevention and control are adhered to. Establish open communication and encourage the patient to express any concerns.

Table 7.6   An overview of the physical examination associated with bladder cancer

Component	Discussion
General examination	Begin with a general physical examination to assess the patient's overall health and vital signs (blood pressure, heart rate, respiratory rate and temperature). Also, observe for any signs of distress or discomfort.
Abdominal examination	The patient is usually positioned in a supine position for an abdominal examination, with arms at the side. This position allows an assessment of the abdomen to be undertaken for any abnormalities, tenderness or masses.
	Gently palpate the abdomen to assess for any tenderness, masses or abnormalities. Pay particular attention to the lower abdomen, as bladder-related issues may manifest in this area.
Genitourinary examination	Perform a focused examination of the genitourinary system, including the external genitalia.
	Inspect for any visible signs of abnormalities, such as lesions or discolouration.
Digital rectal examination	In male patients, a digital rectal examination may be conducted to assess the prostate gland for enlargement or irregularities. Bladder cancer can sometimes involve or affect the prostate.
Pelvic examination	In female patients, a pelvic examination may be performed to assess the reproductive organs and identify any abnormalities in the pelvic region.
Neurological examination	Evaluate neurological function, especially if the patient reports symptoms such as lower back pain or neurological symptoms associated with advanced bladder cancer.
Assessment of lymph nodes	Palpate regional lymph nodes, such as inguinal, iliac and supraclavicular nodes, to check for enlargement. Enlarged lymph nodes may indicate metastasis.
Skin examination	Examine the skin for any signs of metastatic spread, such as nodules or lumps.

Informed consent must be gained before initiating the physical examination, explaining the purpose and the steps involved. Ensure the patient understands the examination process and has the opportunity to ask questions.

Patient privacy is prioritised by providing a private area for undressing and conducting the examination. Use drapes or gowns to cover areas of the patient's body not being examined, promoting a sense of dignity and modesty.

Clearly explain each step of the examination before proceeding. Ensure the patient is positioned comfortably for each aspect of the examination. Maintain clear communication and check for any physical limitations or discomfort that may affect patient positioning.

Table 7.6 provides an outline of the process associated with a physical examination.

STAGING OF BLADDER CANCER

After diagnosis with bladder cancer, it is important to determine if it has spread and, if this is the case, how far. This process is known as staging. Bladder cancer staging describes:

- How far it has grown into the bladder.

- Whether it has spread from where it first started.

Staging helps determine how serious the cancer is and how best to treat it. The stage is one of the most important factors in deciding how to treat the cancer and determining how successful treatment might be. The most commonly used bladder cancer staging system is the TNM system:

Tumour: This explains how far the tumour has grown into the bladder and also how far it has spread into the surrounding tissue.

Nodes: This explains whether the tumour has spread to lymph nodes and is indicated as:

- N0: there is no cancer in the lymph nodes.

- N1: the cancer is in one of the lymph nodes in the pelvis, close to the bladder.

- N2: the cancer is in more than one lymph node within the pelvis.

- N3: there is cancer in one or more of the lymph nodes further away from the bladder but in the abdominal cavity.

Metastasis: This is whether the tumour has spread to another part of the body. Also called secondary or metastatic cancer:

- M0: the cancer has not spread to other parts of the body.

- M1: the cancer has spread to other parts of the body, for example, the bones, lungs, liver, or lymph nodes outside the abdomen.

MUSCLE-INVASIVE BLADDER CANCER STAGING

Non-muscle-invasive bladder cancer means the cancer cells are either in the:

- Inner lining, the urothelium, or

- Connective tissue that surrounds the inner lining of the bladder.

Non-muscle-invasive bladder cancer can be staged as carcinoma in situ (CIS), Ta or T1.

CARCINOMA IN SITU

This is sometimes described as a flat tumour. The cancer cells are only in the inner layer of the bladder lining (urothelium).

TA

The tumour is a mushroom-like growth (papillary cancer). It is only in the inner layer of the bladder lining.

T1

The tumour has started to grow into the layer of connective tissue, beneath the bladder lining. Some people can have both papillary cancer and CIS.

TUMOUR STAGING FOR MUSCLE-INVASIVE AND ADVANCED BLADDER CANCER

Muscle-invasive and advanced bladder cancer is staged as T2–T4.

T2

Tumours have grown into muscle of the bladder wall.

T3

Tumours have grown through the muscle of the bladder and into the fatty tissue around the bladder.

T4

Tumours have spread to other parts of the body near the bladder or in the pelvis.

BLADDER CANCER GRADING

Macmillan Cancer Support (2022) discusses bladder cancer grading. Bladder cancer grading is concerned with how the cancer cells look under a microscope compared with normal cells; it can determine how the cancer might behave. With muscle-invasive or advanced bladder cancer, grading does not affect treatment. With non-muscle-invasive bladder cancer, grading is used to help plan treatment. To determine the grade of cancer cells, tissue samples (biopsies) are taken and analysed by a pathologist (CRUK 2022b). Bladder cancer grades are:

Grade 1: The cancer cells look very similar to normal bladder cells. They are usually slow-growing and are less likely to spread. Generally, the cells stay in the lining of the bladder. It is called low grade or well-differentiated.

Grade 2: The cancer cells look less like normal cells (abnormal) and are slightly faster growing. They are called moderately differentiated. They are more likely to spread into the deeper (muscle) layer of the bladder or to return after treatment.

Grade 3: The cancer cells look very different to normal cells. They are called high grade or poorly differentiated. They grow more quickly and are more likely to return after treatment or spread into the deeper (muscle) layer of the bladder.

Grading bladder cancer is a key component of the diagnostic process that informs management, treatment decisions, guides prognostic assessments and helps to tailor patient management strategies. It plays a pivotal role in the overall understanding of the disease and contributes to personalised and effective cancer care.

INVESTIGATIONS

Making a diagnosis of bladder cancer involves a combination of medical history evaluation, physical examination and various diagnostic investigations. These investigations help in confirming the presence of bladder cancer, determining its stage and grade and guiding the appropriate treatment plan.

Urinalysis is performed to assess for haematuria. Microscopy and culture are undertaken to check for a urine infection. A urine specimen may be tested for cancer cells and substances found in bladder cancer. This is known as molecular testing.

Common diagnostic investigations for bladder cancer are outlined in Table 7.7.

The combination of the investigations outlined in Table 7.7 can help to establish a comprehensive understanding of the extent and characteristics of bladder cancer. The results help to guide the staging and grading of the cancer, which, in turn, will inform the development of an appropriate and personalised treatment plan. Early detection and accurate diagnosis are crucial for promoting positive outcomes and determining the most effective course of treatment for bladder cancer.

Table 7.7 Common diagnostic investigations for bladder cancer

Investigation	Discussion
Urinalysis	Helps detect haematuria.
Cystoscopy	This procedure permits visual inspection of the interior of the bladder, to enable the identification of any abnormalities and to take biopsies for further examination.
Imaging studies	• Ultrasound: Transabdominal or transrectal ultrasound may be used to visualise the bladder and surrounding structures. • Computed tomography (CT) urography: A CT scan with contrast can provide detailed images of the urinary tract, including the bladder and nearby lymph nodes. • MRI urography: An MRI may be used to obtain detailed images of the bladder and surrounding structures.
Urine cytology	This test involves examining a sample of urine under a microscope to identify abnormal cells shed by the bladder lining. While not as sensitive as cystoscopy, it can provide additional information.
Biopsy	Tissue biopsy is crucial for confirming the diagnosis and determining the type and grade of bladder cancer. Biopsy samples are obtained during cystoscopy and analysed by a pathologist.
Urodynamic studies	In cases where there are concerns about how well the bladder is functioning, urodynamic studies may be conducted to assess bladder and urethral function.
Blood tests	Blood tests may be performed to assess kidney function and overall health. In some cases, specific markers (bio markers) are associated with bladder cancer.
Voiding cystourethrogram	This imaging study involves injecting contrast dye into the bladder to visualise the anatomy and assessing for any abnormalities in the urethra and bladder.
Bone scan	In cases where there is a suspicion of advanced disease and potential metastasis, a bone scan may be performed to check for the spread of cancer to the bones.
Chest X-ray	A chest X-ray may be ordered to assess the lungs and check for any signs of metastasis.

MANAGEMENT

The National Institute for Health and Care Excellence (2015) notes that the involvement of the urogenital tract and the nature of the treatments means that bladder cancer has a strong psychological impact, as well as the physical impact of the disease and its treatments, which are often profound.

Bladder cancer is highly treatable when it is diagnosed in the early stages. Depending on the stage and grade of the cancer, the type of cancer, the number of tumours and other factors, treatment options for people with bladder cancer can include:

- Bladder cancer surgery

- Intravesical therapy

- Chemotherapy

- Radiation therapy

- Immunotherapy

- Targeted therapy drugs

TREATING NON-MUSCLE-INVASIVE BLADDER CANCER

Non-muscle-invasive bladder cancer means the cancer cells are in the inner lining or the connective tissue surrounding the inner lining of the bladder. There is no spread into the muscle layer below this. Treatments may include the following:

Transurethral resection of a bladder tumour (TURBT): A TURBT is a surgical operation to remove bladder cancer from the inside of the bladder. It is the main treatment for non-muscle-invasive bladder cancer.

Chemotherapy: Uses cytotoxic drugs to destroy cancer cells. Intravesical chemotherapy is when the chemotherapy is given directly into the bladder through a catheter.

Bacillus Calmette–Guérin (BCG): A type of immunotherapy drug. Immunotherapy drugs use the body's immune system to find and attack cancer cells. For bladder cancer, BCG is given intravesically.

Surgery: Some people may have a cystectomy if they have:

- A high-risk cancer

- A cancer that keeps coming back and treatments are not working

There are sometimes other treatments that may be used to treat non-muscle-invasive bladder cancer. Some of the treatments may only be available at certain hospitals, for example, tumour ablation, electromotive intravesical chemotherapy or heated intravesical chemotherapy.

TREATING MUSCLE-INVASIVE BLADDER CANCER

Muscle-invasive means that the cancer has spread into or through the muscle layer of the bladder wall. Treatment options aim to cure the cancer with one of the following:

Cystectomy: It is surgery to remove the bladder and create urinary diversion. Intravenous chemotherapy may be administered before or after the operation.

Radical radiotherapy: High-energy rays are used in radical radiotherapy to destroy cancer cells. It is a radiotherapy that is used with other treatments, for example, surgery or chemotherapy.

The choice of treatment is personalised, working closely with the patient to determine the most appropriate approach based on the specific characteristics of the cancer and the individual's overall health. It is important for patients to actively participate in the decision-making process. Regardless of the treatment modality, regular follow-up care is essential to monitor for recurrence and manage potential side effects. Follow-up may include imaging studies, cystoscopy and other tests to ensure early detection of any signs of cancer recurrence.

HEALTH TEACHING

Health teaching for individuals regarding bladder cancer involves providing information and guidance to help them understand the condition, its risk factors, symptoms, prevention strategies and the importance of early detection.

A crucial aspect is informing individuals about the risk factors and preventive measures associated with bladder cancer. This includes highlighting the strong link between smoking and bladder cancer, advocating for smoking cessation and providing information about potential occupational exposures to chemicals and dyes that may increase the risk.

Early detection plays a key role in managing bladder cancer. Educating individuals about early signs and symptoms, particularly the significance of haematuria, is crucial. Encouraging prompt medical attention for any unusual changes in urine colour or urinary symptoms helps in early diagnosis. Regular health check-ups, especially for those with risk factors, are recommended and the importance of discussing any concerns with healthcare providers is emphasised.

Promoting a healthy lifestyle is another key component of health teaching. This involves advocating for a balanced diet. Staying hydrated through adequate water intake is encouraged, as it may help dilute potentially harmful substances in the urine. Regular physical activity is also promoted for its overall health benefits and potential risk reduction.

For individuals who have undergone treatment, the importance of regular follow-up appointments and monitoring for signs of recurrence is highlighted. Adherence to medical recommendations, including completing prescribed treatments and medications, is stressed for optimal outcomes.

Psychosocial support is recognised as integral to the well-being of individuals dealing with bladder cancer. Addressing the potential emotional impact of a diagnosis and encouraging individuals to seek support from friends, family, or support groups is emphasised. Information about available counselling services or mental health support is provided to help individuals cope with the challenges associated with bladder cancer.

Addressing the impact of bladder cancer and its treatment on sexual health is an important aspect of health teaching. Open communication is required to address concerns, and considerations related to sexual function are encouraged.

This approach to health teaching aims to empower individuals, enhance their awareness and promote proactive engagement in their healthcare journey. Effective health teaching empowers individuals to make informed decisions about their health, adopt preventive measures and actively participate in their healthcare. Tailoring information to individual risk factors and addressing concerns in a supportive manner can contribute to better awareness and outcomes for bladder cancer.

CONCLUSION

This chapter has offered insight into the intricate landscape of bladder cancer, exploring its pathophysiology, risk factors, diagnostic approaches and varied treatment modalities. Bladder cancer, characterised by its heterogeneity, poses a multifaceted challenge that necessitates a nuanced understanding for effective management.

The pathophysiological journey of bladder cancer, from the initiation of genetic alterations to the progression into invasive stages, highlights the complexity of the disease. Diagnostic investigations, ranging from urinalysis and cystoscopy to advanced imaging and biopsies, are essential tools that can be used to ensure care provision is patient-centred. Early detection remains paramount, and the various diagnostic tools play a pivotal role in determining the disease's stage, grade and potential spread.

Treatment options, tailored to the specific characteristics of the cancer and the individual patient, span from localised interventions such as transurethral resection and intravesical therapy to more extensive measures such as radical cystectomy for advanced cases.

Moreover, health teaching has been stressed as an integral component in the holistic management of bladder cancer. Empowering individuals with knowledge about risk factors, preventive measures and the importance of early detection fosters active participation in their health journey. Additionally, addressing psychosocial aspects, promoting healthy lifestyles and providing ongoing support contribute to a comprehensive approach to patient care.

GLOSSARY OF TERMS

Biopsy: Removal and examination of a small tissue sample to determine the presence of cancer cells.

Carcinogen: A substance or agent that can cause cancer.

Carcinoma: A type of cancer that starts in epithelial cells lining organs, including the bladder.

Chemotherapy: Treatment using drugs to destroy cancer cells or slow their growth.

Cystectomy: Surgical removal of the bladder.

Cystoscopy: A procedure using a thin tube with a camera to examine the inside of the bladder.

Immunotherapy: Treatment that stimulates the body's immune system to fight cancer cells.

Intravesical therapy: Treatment involves delivering medication directly into the bladder.

Metastasis: Spread of cancer from one organ or part of the body to another not directly connected with it.

Muscle-invasive bladder cancer: A type of bladder cancer that has penetrated the muscular layers of the bladder wall.

Non-muscle-invasive bladder cancer: Cancer that has not spread beyond the inner lining of the bladder.

Pathology: Study of diseases, including examining tissues and cells for diagnosis.

Radiation therapy: Treatment using high-dose radiation to kill or damage cancer cells.

Staging: Determining the extent of cancer spread in the body.

Transurethral resection of bladder tumour: Surgical procedure to remove tumours from the bladder lining.

Tumour: Abnormal mass of tissue formed by the uncontrolled growth of cells.

MULTIPLE CHOICE QUESTIONS

1. Which type of bladder cancer is characterised by the invasion of cancer cells into the muscular layers of the bladder wall?
   a) Non-muscle-invasive bladder cancer
   b) Muscle-invasive bladder cancer
   c) Papillary bladder cancer
   d) Superficial bladder cancer

2. What is the most common symptom of bladder cancer?
   a) Headache
   b) Haematuria
   c) Shortness of breath
   d) Joint pain

3. What diagnostic procedure is commonly used to visualise the inside of the bladder for signs of cancer?
   a) Colonoscopy
   b) Bronchoscopy
   c) Cystoscopy
   d) Arthroscopy

4. Which of the following is a risk factor for developing bladder cancer?
   a) Regular exercise
   b) Smoking
   c) Adequate hydration
   d) Vegetarian diet

5. What is the primary treatment for non-muscle-invasive bladder cancer (NMIBC)?
   a) Radical cystectomy
   b) Chemotherapy
   c) Transurethral resection of bladder tumour
   d) Radiation therapy

6. What is the primary side effect of intravesical therapy for bladder cancer?
   a) Hair loss
   b) Nausea
   c) Bladder irritation
   d) Fatigue

7. What is the surgical procedure for the removal of the bladder called?
   a) Hysterectomy
   b) Colectomy
   c) Cystectomy
   d) Nephrectomy

8. What is the term for the spread of cancer from one organ to another not directly connected with it?
   a) Metastasis
   b) Infiltration
   c) Proliferation
   d) Differentiation

9. Which term refers to the study of diseases, including examining tissues and cells for diagnosis?
   a) Epidemiology
   b) Histology
   c) Radiology
   d) Pathology

10. What type of treatment stimulates the body's immune system to fight cancer cells?
    a) Chemotherapy
    b) Immunotherapy
    c) Hormone therapy
    d) Radiation therapy

REFERENCES

Cancer Research UK (2022a). Symptoms of bladder cancer. https://www.cancerresearchuk.org/about-cancer/bladder-cancer/symptoms (accessed December 2023).

Cancer Research UK (2022b). Grades of bladder cancer. https://www.cancerresearchuk.org/about-cancer/bladder-cancer (accessed December 2023).

Cancer Research UK (2023). Risks and causes of bladder cancer. https://www.cancerresearchuk.org/about-cancer/bladder-cancer/risks-causes (accessed December 2023).

Centers for Disease Control and Prevention (2023). Bladder cancer. https://www.cdc.gov/cancer/bladder/index.htm (accessed December 2023).

Leach, D. (2023). Urological cancer (Chapter 8). In: *Stoma Care Specialist Nursing: A Guide for Clinical Practice* (ed. M. White and A. Perrin). New York: Springer International Publishing AG.

Macmillan Cancer Support (2022). Staging and grading of bladder cancer. https://www.macmillan.org.uk/cancer-information-and-support/bladder-cancer/staging-and-grading-of-bladder-cancer (accessed December 2023).

National Cancer Institute (2023). Bladder cancer cause and risk factors. https://www.cancer.gov/types/bladder/causes-risk-factors (accessed December 2023).

National Institute for Health and Care Excellence (2015). Bladder cancer: diagnosis and management. https://www.nice.org.uk/guidance/ng2/resources/bladder-cancer-diagnosis-and-management-pdf-51036766405 (accessed December 2023).

National Institute for Health and Care Excellence (2021). URO17 for detecting bladder cancer. https://www.nice.org.uk/advice/mib250/resources/uro17-for-detecting-bladder-cancer-pdf-2285965686694597 (accessed December 2023).

Senior, T. (2021). The physical examination. *British Journal of General Practice* 71 (709): 372. doi: 10.3399/bjgp21X71672.

 # Urinary Tract Infection

A urinary tract infection is a microbial infection that can affect any part of the urinary system, including the kidneys, bladder, ureters and urethra. The majority of urinary tract infections are caused by bacteria, with *Escherichia coli* being the most common pathogen (Yaqoob and Ashman 2021). The National Institute for Health and Care Excellence [NICE] 2018) defines a recurrent urinary tract infection as two proven episodes within six months or three within a year.

Urinary tract infections can be divided into upper tract infections, which involve the kidneys (pyelonephritis) and lower tract infections, involving the bladder (cystitis), urethra (urethritis) and prostate (prostatitis). It may, in practice, be a challenge to differentiate between the sites. Furthermore, infection can often spread from one area to the other.

Chapter 3 of this book discusses pyelonephritis. This chapter will focus on lower urinary tract infections. In Table 8.1, the European Association of Urology ([EAU] 2023) provides classifications of urinary tract infections.

Table 8.1 European Association of Urology classification of urinary tract infections

Classification	Discussion
Uncomplicated urinary tract infections	Acute, sporadic or recurrent lower (uncomplicated cystitis) and/or upper (uncomplicated pyelonephritis) urinary tract infection, limited to non-pregnant women with no known relevant anatomical and functional abnormalities within the urinary tract or comorbidities.
Complicated urinary tract infections	All urinary tract infections which are not defined as uncomplicated. Meaning in a narrower sense, urinary tract infections in a patient with an increased chance of a complicated course, i.e. all men, pregnant women, patients with relevant anatomical or functional abnormalities of the urinary tract, an indwelling urinary catheter, renal diseases and/or with other related immunocompromising diseases, for example, diabetes.
Recurrent urinary tract infections	Recurrences of uncomplicated and/or complicated urinary tract infections, with a frequency of at least three urinary tract infections per year or two urinary tract infections in the last six months.
Catheter-associated urinary tract infections	Catheter-associated urinary tract infection refers to urinary tract infections occurring in a person whose urinary tract is currently catheterised or has had a catheter in place within the past 48 hours.
Urosepsis	Urosepsis is a serious condition where the body's organs start malfunctioning because of an uncontrolled reaction to an infection that comes from the urinary tract or male genital organs. It is a potentially life-threatening situation that results from the body's response to the infection rather than the infection itself.

Source: Adapted from EAU (2023).

PATHOPHYSIOLOGICAL CHANGES ASSOCIATED WITH URINARY TRACT INFECTION

Infections of the urinary tract represent major public health problems around the world. The most common reasons for infections are either functional (for example, problems with the way urine is transported, stored or eliminated) or anatomical abnormalities (for example, physical irregularities or malformations in the kidneys, ureters, bladder or urethra). These infections are common in all individuals: men, women, children or the elderly. This occurs as the urethra not only serves as a passage for the release of urine but also as an entrance for the bacteria into the urinary tract (Kaur and Kaur 2020).

Typically, bacteria reside in the vicinity of the urinary tract opening (urethral meatus), forming colonies in both men and women. Normally, they are expelled during urination. Infections arise when these colonies persist and reach the bladder before urination. Women are more prone to urinary tract infections due to a shorter distance between the urethral opening and the bladder (Yaqoob and Ashman 2021). The likelihood of bacterial colonies forming is increased in women due to the presence of the vaginal cavity and the close proximity of the rectum to the urethral opening. Infections typically arise due to bacteria originating from a patient's own intestinal flora (Yaqoob and Ashman 2021). Notably, even if these bacteria reach the bladder and multiply, urinary tract infection symptoms are often not immediately apparent.

Urinary tract infections involve the invasion and multiplication of microorganisms, typically bacteria, within the urinary tract. Other microorganisms can invade the urinary tract. The most common non-bacterial pathogens are fungi (usually candida species) and less commonly, mycobacteria, viruses and parasites (Hagler et al. 2023). Non-bacterial pathogens usually affect patients who are immunocompromised; have diabetes, obstruction or structural urinary tract abnormalities; or have had recent urinary tract instrumentation (for example, cystoscopy).

The pathophysiological changes associated with urinary tract infections can vary depending on the location and severity of the infection.

Urinary tract infections often commence with the entry of bacteria into the urinary tract. Bacteria enter through the urethra and ascend into the bladder. Following bacterial entry, the urethra may become colonised. Depending on factors such as sexual activity, catheterisation or conditions affecting normal urinary tract function, bacteria may ascend further into the bladder.

When bacteria ascend and reach the bladder, it can lead to cystitis, resulting in inflammation due to the body's immune response. Symptoms include increased frequency and urgency of urination, pain or discomfort during urination and lower abdominal discomfort.

In more severe cases, bacteria may ascend to infect the kidneys, which can cause pyelonephritis. This condition can lead to systemic symptoms such as fever, chills, nausea and back or flank pain. The infection may damage renal tissues and impair kidney function (see Chapter 3).

The presence of bacteria triggers an inflammatory response by the immune system, involving the release of inflammatory mediators, recruitment of immune cells and activation of various immune pathways. Inflammation contributes to symptoms such as redness, swelling and pain.

Bacterial invasion and the host's inflammatory response can cause damage to the mucosal lining of the urinary tract. This can lead to potential bleeding and contribute to symptoms of haematuria.

In severe cases, particularly with kidney involvement, bacteria and inflammatory products may enter the bloodstream, resulting in bacteraemia. This can lead to systemic effects such as fever, malaise and, in extreme cases, sepsis.

EPIDEMIOLOGY

Urinary tract infection is the most common bacterial infection overall for men and women in all patient-care settings. Urinary tract infection is much less common in men (who account for 20% of all occurrences) than in women – this is attributed to the shorter urethra. The incidence of urinary tract infection in men increases substantially with:

- Increasing age (these men are more likely to have additional risk factors): urinary tract infection is very uncommon in otherwise healthy young and middle-aged men and rarely develops in men before 50 years of age.

- Institutional care: residence in a long-term care facility correlates with the likelihood of men developing bacteriuria (the presence of bacteria in the urine) and urinary tract infection.

- An indwelling urinary catheter.

Acute urinary tract infection occurs in up to 50% of women (Yaqoob and Ashman 2021). Estimates suggest that by the age of 24 years, nearly one-third of females will have had at least one episode of cystitis. Recurrence occurs in 20–30% of women who have had a urinary tract infection. Urinary tract infections are common in older women.

Urinary tract infection is the second most common healthcare-associated infection (after respiratory tract infection), accounting for 17.2% of all healthcare-associated infections.

- Up to 50% of these occur in people with a urinary catheter.

Asymptomatic bacteriuria is more common in older people with comorbidities, in institutional care and/or with an indwelling urinary catheter. The prevalence of asymptomatic bacteriuria is:

- Up to 10% in men aged over 80 years in the community.

- Up to 40% in institutionalised men.

Bacteriuria develops within days of catheter insertion and over time all people with a catheter have bacteriuria. The prevalence of catheter-associated urinary tract infections is estimated to be 8.5%. Approximately half of healthcare-acquired infections are due to an indwelling urinary catheter. Catheter-associated urinary tract infections are one of the main causes of secondary healthcare-associated bacteraemia (the presence of bacteria in the bloodstream; NICE 2023a,b; EAU 2023).

RISK FACTORS

Urinary tract infections are influenced by various risk factors that can contribute to the introduction and proliferation of bacteria in the urinary system. See Table 8.2 for a summary of risk factors.

Table 8.2 Summary of risk factors associated with urinary tract infection

Risk factors	Discussion
Gender	Women are more prone to urinary tract infections than men.
Age	Elderly individuals may have an increased risk of urinary tract infections due to factors such as weakened immune function, incomplete emptying of the bladder or the presence of underlying health conditions.
Urinary tract abnormalities	Structural abnormalities in the urinary tract, such as renal calculi (see Chapter 6), urinary retention or urethral strictures, can create pockets where bacteria can accumulate, increasing the risk of infection. Benign prostatic hypertrophy can result in urine outflow obstruction.
Catheter use	Individuals who use urinary catheters, often due to medical conditions or after surgery, have an elevated risk of urinary tract infections. Catheters provide a pathway for bacteria to enter the bladder.
Sexual activity	Women who are sexually active are at a higher risk of developing urinary tract infections. Activities such as sexual intercourse can introduce bacteria into the urethra and contribute to infection. A new sex partner within the last year increases risk.
Menopause	The hormonal changes associated with menopause can alter the urinary tract environment, making women more susceptible to urinary tract infections.
Pregnancy	Pregnant women may experience changes in the urinary tract and the growing uterus can exert pressure on the bladder, reducing its ability to empty completely.
Immunocompromised	Conditions that weaken the immune system, such as diabetes, HIV/AIDS or undergoing chemotherapy, increase the vulnerability to infections, including urinary tract infections.
Genetic factors	Some individuals may be genetically predisposed to recurrent urinary tract infections.
Voiding habits	Holding in urine for extended periods or not fully emptying the bladder during urination can create conditions conducive to bacterial growth.
Use of spermicides and diaphragms	Certain contraceptive methods, such as spermicides and diaphragms, may alter the bacterial balance in the vagina and increase the risk of urinary tract infections.
Previous urinary tract infections	Having a history of urinary tract infections increases the likelihood of future occurrences.
Dementia	Dementia and urinary tract infections can be interconnected; several factors contribute to this association: • Communication challenges • Impaired hygiene practices • Immune system compromise • Reduced mobility • Incontinence • Dehydration • Catheter use • Difficulty in recognising and seeking help

Source: Adapted from O'Callaghan (2016), Yaqoob and Ashman (2021) and Hagler et al. (2023).

CLINICAL PRESENTATION

Hagler et al. (2023) note that manifestations of urinary tract infections can range from painful urination associated with uncomplicated urethritis or cystitis to severe systemic illness accompanied by abdominal or flank pain, pyrexia or sepsis. The most typical symptoms of (lower) urinary tract infection are (Swales 2022):

- Frequency of micturition by day and night

- Urgency

- Dysuria

- Suprapubic pain and tenderness

- Haematuria

- Cloudy urine

- Malodorous urine

- Incontinence

- Malaise

- Worsening delirium/debility

These are the symptoms of bladder and urethral inflammation or 'cystitis'. Loin pain and tenderness, with pyrexia, chills, night sweats and rigours, would suggest there has been an extension of infection to the renal pelvis and kidney, known as pyelitis or pyelonephritis. Determining the specific location of an infection solely based on the observed symptoms is not a reliable or accurate method (Yaqoob and Ashman 2021). It is not always possible to pinpoint the exact site of infection in the body just by considering the symptoms a person is experiencing. See Figure 8.1 for urinary tract infection.

Urinary tract infections can also present with few or no symptoms (particularly in those who are immunocompromised) or may manifest as abdominal pain, pyrexia or haematuria in the absence of frequency or dysuria. In the elderly, new confusion may be the only symptom of urinary tract infection.

CLINICAL INVESTIGATIONS AND DIAGNOSIS

To make a diagnosis of urinary tract infection, a combination of clinical assessments, patient history and laboratory tests is required to ensure accuracy. Urinary tract infections are detected by analysis of bacteria in the urine of the patient (Kaur and Kaur 2020). Efforts are being made for the development of rapid detection approaches, monitoring and quantification of uropathogens (microorganisms capable of causing urinary tract infections). When left undetected in the early stages, urinary tract infections can cause serious health implications.

HISTORY TAKING

The initial step in making a diagnosis includes gathering a detailed patient history. During the history-taking phase, inquiries are made about symptoms and relevant medical history. History is a crucial component in diagnosing urinary tract infections as it helps gather valuable information about a patient's symptoms, risk factors and overall health. It is important to

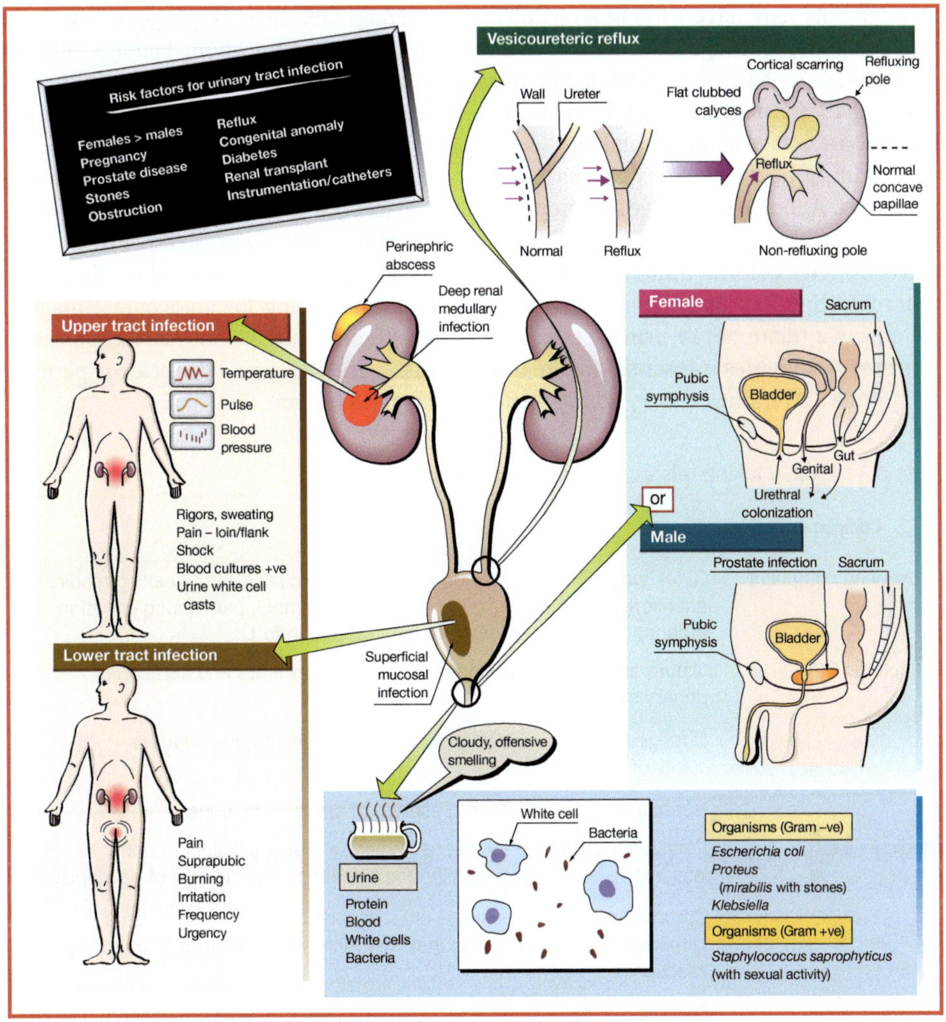

FIGURE 8.1   Urinary tract infection

recognise some of the sensitivities associated with the history-taking process. Understanding and addressing these sensitivities can enhance the effectiveness of history taking and contribute to a more patient-centred approach.

Discussing urinary symptoms and sexual history can be sensitive. Healthcare providers should create a private and comfortable environment, use clear and non-judgmental language and ensure patients feel at ease. Be mindful that patient perspectives on health and illness can vary based on cultural backgrounds, be aware of cultural differences, show respect and tailor the approach to each individual's cultural norms.

Patients may have a gender preference for discussing intimate topics. Offering patients the option to choose the gender of their healthcare provider can enhance comfort during sensitive discussions. There may be patients who have a history of trauma, abuse or experiences that affect how comfortable they feel discussing certain topics; therefore, creating a safe environment and allowing patients to share information at their own pace is important.

Patients may have varying levels of health literacy. Use clear language, assess understanding and encourage questions to ensure effective communication. Patients may fear judgement or embarrassment related to urinary tract infections, especially due to misconceptions. Offer information and advice on urinary tract infections and dispel myths, emphasising that they are common and treatable.

Urinary tract infections can have emotional effects, especially if they recur or cause discomfort. Adopt a supportive and empathetic approach, acknowledge the emotional impact of urinary tract infections and address any concerns or anxieties. Approaching history taking with sensitivity, cultural awareness and respect for the patient's comfort level facilitates open communication, builds trust and enables information gathering for an accurate diagnosis and effective treatment of urinary tract infections. Patient-centred care that recognises and addresses sensitivities is essential for creating a positive and inclusive healthcare experience.

Table 8.3 outlines history taking related to urinary tract infection.

Table 8.3 History taking related to urinary tract infection

Component	Discussion
Presenting complaint	Chief complaint: Determine the patient's main reason for seeking medical attention, such as urinary symptoms, for example, pain during urination, urgency, frequency or lower abdominal discomfort.
Symptoms	Nature and duration: understand the characteristics and duration of urinary symptoms.
	Changes in urine: inquire about changes in the colour, odour or clarity of urine.
	Haematuria: ask about the presence of blood in the urine.
Medical history	Previous urinary tract infections: inquire about any history of urinary tract infections, urolithiasis, including the frequency and severity of past episodes.
	Chronic illnesses: identify any chronic conditions such as diabetes or immunocompromised states that may increase susceptibility to urinary tract infections.
	Determine if there is any cognitive impairment (new or established); urinary tract infection may present atypically with delirium or debility.
Family history	Family history of urinary tract disease such as polycystic kidney disease.
Medication history	Current medications: identify medications that the patient is currently taking, as some drugs can impact urinary function or increase the risk of urinary tract infections.
Sexual history	Sexual activity: inquire about sexual activity, as sexual intercourse can be a risk factor for urinary tract infections, especially in women.
Reproductive history	Menstrual and menopausal history: in women, understanding the menstrual cycle and menopausal status can provide relevant information.
	Pregnancy: is there a possibility of pregnancy?
Voiding habits	Frequency and urgency: explore patterns of urination, including how often the patient voids and whether there is a sense of urgency.
	Voiding position: inquire about any difficulties or changes in the voiding position.
Hygiene practices	Perineal hygiene: discuss perineal hygiene practices, as poor hygiene can contribute to urinary tract infections.

Component	Discussion
Social history	Lifestyle factors: consider lifestyle factors such as fluid intake, diet and occupation.
Allergies	Known allergies: identify any known allergies, especially to medications that might be prescribed for urinary tract infection treatment.

THE PHYSICAL EXAMINATION

The physical examination assesses the patient's overall health, identifies specific signs of a urinary tract infection and can help to rule out other potential causes of urinary symptoms. As is the case with history taking, it is important to recognise some of the sensitivities associated with a physical examination. Understanding and addressing these sensitivities can enhance the effectiveness of history taking and contribute to a more patient-centred approach. The patient should be offered a chaperone during the examination. Local policy and procedure must be adhered to at all times in regard to infection prevention and control. Informed consent must be obtained.

It should be noted that the specific components of the physical examination may vary based on the patient's gender, age and clinical presentation.

General Examination

Vital signs: Check vital signs, including temperature, blood pressure, heart rate and respiratory rate, to assess for signs of systemic infection. Observe the patient for signs of wellness and for any indications of systemic illness. Note if the patient has a urinary catheter in situ.

Abdominal Examination

Palpation: Palpating the abdomen to assess for tenderness or pain, which may indicate inflammation or infection in the bladder or kidneys.

Costovertebral angle tenderness: Evaluate for tenderness over the costovertebral angle, which may suggest kidney involvement (see also Chapter 3).

Genitourinary Examination

External genitalia inspection: Inspect the external genitalia for signs of irritation, redness or discharge.

Urethral meatus examination: Check the urethral meatus for signs of inflammation or discharge.

Pelvic Examination (for Women)

Speculum examination: Conduct a speculum examination of the vagina to assess for any signs of infection or irritation.

Cervical examination: Check the cervix for tenderness or discharge.

Digital Rectal Examination (for Men)

Prostate examination: Conduct a digital rectal examination to assess the prostate for enlargement, tenderness or abnormalities.

Neurological Examination

Assessing for neurological symptoms: Inquire about and assessing for any neurological symptoms that may be associated with urinary dysfunction.

Skin Examination

Skin inspection: Inspect the skin for any signs of infection or lesions that may suggest the spread of infection.

Musculoskeletal Examination

Assessing mobility: Inquire about and assess the patient's mobility, which may be relevant if musculoskeletal conditions are contributing to urinary symptoms.

Additional diagnostic tests such as urinalysis, urine culture and imaging studies may also be required to confirm the diagnosis and guide treatment. The combination of a thorough history, physical examination and appropriate diagnostic tests can enable an accurate diagnosis and develop an effective treatment plan for people with urinary tract infections.

CLINICAL INVESTIGATIONS

Recommendations and guidelines regarding the investigations required for urinary tract infection have been made by EAU (2023), Public Health England ([PHE] 2020) and NICE (2023a,b,c).

One of the primary diagnostic tools is urinalysis, which encompasses a urine dipstick test and microscopic examination. The dipstick test checks for indicators such as white blood cells, red blood cells, nitrites and bacteria. Nitrites are particularly significant as they are often produced by certain bacteria that cause urinary tract infections. Microscopic examination involves examining a urine sample under a microscope to identify abnormalities such as white blood cells, red blood cells or bacteria.

OBTAINING A MIDSTREAM SPECIMEN OF URINE

A midstream specimen of urine sample is the standard for urine sampling. Clare (2021) discusses midstream specimens of urine sample collection. Using the midstream sample method intends to reduce the chances of contamination of the specimen. The midstream allows for debris and cutaneous bacteria to be flushed away prior to the sample being taken, avoiding contamination. The procedure for obtaining a midstream specimen of urine sample requires the full cooperation of the patient and is usually undertaken by the patient themselves. If a patient is unable to follow instructions, then obtaining a midstream sample is not possible and this should be noted on the sample form, so the laboratory is aware.

The procedure is explained to the patient on how to take a midstream specimen of urine:

- Provide the patient with sterile urine specimen collection equipment and explain the procedure to them.

- Obtain consent for testing of the sample.

- Ensure that the patient has a private place to perform collection of the specimen. If not in a locked toilet cubicle, then ensure that the curtains are drawn around the bed and no one will enter.

- Ask the patient to wash their hands.

- The patient should then retract the foreskin (if present in males) or retract the labia and wash the surrounding area with 0.9% saline or disinfectant-free soap and water and then allow it to air dry. This is done to reduce the possibility of contamination of the mucosa by skin-dwelling bacteria. Products containing disinfectant should be avoided, as they may irritate the skin.

- The patient should then pass the first third of the urine into the toilet or other receptacle. This is intended to 'flush' away any transient bacteria in the urethra.

- Without interrupting the flow, the client should introduce the wide-necked collection equipment into the flow and collect a sample of approximately 15–30 mL.

- The remainder of the urine can then be passed into the toilet (or other receptacle).

- The patient should then wash their hands.

Once the specimen has been obtained, the sample can then be sealed, labelled and placed into the appropriate specimen bag. Local policy and procedure must be adhered to at all times (see Figure 8.2).

A urine culture can also be used to determine the specific bacteria responsible for the infection. This not only aids in confirming the diagnosis but also guides the selection of the most appropriate antibiotic treatment. Swales (2022) notes that urine culture is required for high-risk patients, such as those who are pregnant or immunosuppressed or those who have failed to respond to earlier antibiotic treatment. Urine culture should always be undertaken in men with a history suggestive of urinary tract infection, regardless of the results of the urinalysis.

Blood tests, such as a full blood count, and assessments of inflammatory markers, for example, C-reactive protein or erythrocyte sedimentation rate, are conducted to evaluate the presence of infection and inflammation.

In cases where there may be structural abnormalities or kidney involvement is suspected, imaging studies such as ultrasound or CT scans may be recommended. Ultrasound allows for visualisation of the kidneys and bladder and CT scans can offer more detailed imaging, especially in complicated cases. In some instances, cystoscopy may be employed for a direct visual examination of the urethra and bladder with the intention of identifying any abnormalities.

When used collectively, these diagnostic tools contribute to a comprehensive assessment, enabling confirmation of the diagnosis, determining the severity of the infection, identifying the causative bacteria and tailoring appropriate treatment strategies for individuals with urinary tract infections.

MANAGEMENT

The main objective is to identify the cause, offer appropriate treatment so as to irradicate it (Jones and Steggall 2019) and prevent upper urinary tract involvement. The patient will need reassurance, psychological support and health education. It is important to report signs of groin pain, restlessness, sweating and tachycardia for prompt action to be taken. These symptoms may indicate the possibility of a more severe or complicated infection that has extended

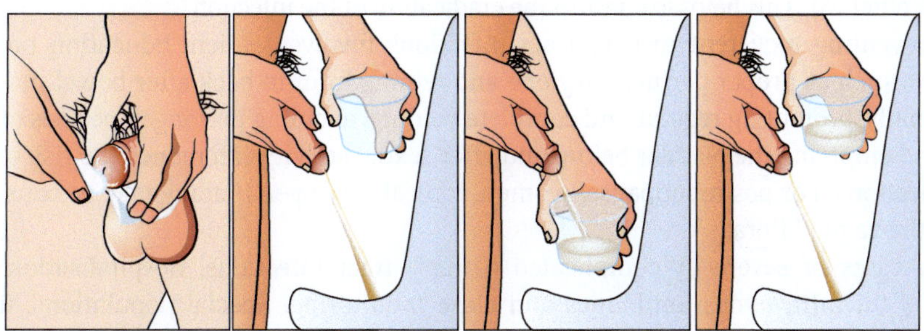

FIGURE 8.2   Obtaining a midstream specimen of urine from a male patient

beyond the lower urinary tract. The management of urinary tract infections can involve a combination of antimicrobial therapy, supportive care (including pain management) and measures to prevent recurrent infections.

A comprehensive approach is advocated in order to address symptoms, eradicate the infection and prevent recurrence. When patients present with symptoms such as pain during urination, increased frequency and discomfort, it is usual to initiate empirical antibiotic treatment that is based on the likely pathogens that are causing the urinary tract infection. Hagler et al. (2013) suggest that in cases of uncomplicated urinary infection, a short-term dose of antibiotic is provided. In some cases, urine cultures are obtained to identify the specific bacteria and this then guides the selection of the appropriate antimicrobial therapy.

There are a range of commonly prescribed antibiotics for uncomplicated urinary tract infections, these include:

- Trimethoprim

- Nitrofurantoin

- Pivmecillinam

- Fosfomycin

- Amoxicillin

- Cefalexin

It is crucial to identify any allergies during the history taking phase of patient assessment, particularly when considering the prescription of antibiotics. Allergies can significantly impact the choice of antibiotics, as certain individuals may have adverse reactions or hypersensitivity to specific antibiotic classes. If a patient has a documented allergy to a particular antibiotic, prescribing that antibiotic can lead to serious and potentially life-threatening allergic reactions. An alternative antibiotic not associated with the allergic reaction will need to be prescribed. Identifying allergies ensures that the prescribed antibiotics align with the patient's medical history, minimises the risk of adverse reactions and contributes to successful treatment outcomes.

For those patients with a fungal-induced urinary tract infection, antifungal medication is required, for example, fluconazole (Hagler et al. 2023).

The choice of antibiotic may vary for special populations, such as pregnant women or individuals with allergies, and local prescribing guidance is advised. Supportive measures include encouraging increased fluid intake to flush bacteria from the urinary tract and managing pain and discomfort (such as pyrexia) with analgesics/antipyretics (Swales 2022).

Patients are monitored for the resolution of symptoms and may be provided with follow-up appointments to assess the effectiveness of antibiotic therapy. In certain cases, repeat urine cultures are obtained. This helps to confirm the eradication of the infection.

Preventing recurrent urinary tract infections involves patient education on hygiene practices, such as proper perineal hygiene and wiping front to back after bowel movements. Additionally, promoting regular and complete voiding, avoiding prolonged periods of holding urine and emptying the bladder before and after sexual activity can reduce the risk of urinary tract infections. For postmenopausal women, topical oestrogen therapy may be considered to restore the vaginal flora.

In cases of severe or complicated urinary tract infections, hospitalisation may be necessary for intravenous antibiotics and close monitoring. Special populations, including pregnant women, individuals with diabetes and those with anatomical abnormalities, require individualised management approaches.

Table 8.4 An overview of the key aspects of managing people with urinary tract infection

Antimicrobial therapy	Empirical treatment: Empirical antibiotic treatment based on the likely pathogens causing urinary tract infection.
	Culture-guided therapy: Urine cultures identify the specific bacteria and determine their susceptibility to antibiotics, guiding the selection of appropriate antimicrobial therapy.
Choice of antibiotics	First-line antibiotics: Commonly prescribed antibiotics for uncomplicated urinary tract infections include trimethoprim, nitrofurantoin and fluoroquinolones.
	Pregnant women and special populations: Antibiotic choice may vary for pregnant women, individuals with allergies or those with complicating factors.
Supportive measures	Fluid intake: Encouraging increased fluid intake helps flush bacteria from the urinary tract.
	Pain management: Analgesics such as paracetamol or non-steroidal anti-inflammatory drugs can help manage pain and discomfort.
Follow-up and monitoring	Symptom resolution: Patients are monitored for resolution of symptoms and a follow-up appointment may be scheduled to ensure the effectiveness of antibiotic therapy.
	Repeat cultures: Repeat urine cultures may be obtained to confirm the eradication of the infection in some cases.
Prevention of recurrent urinary tract infections	Hygiene practices: Advising patients on proper perineal hygiene and wiping front to back after bowel movements can help prevent the introduction of bacteria into the urethra.
	Voiding habits: Encouraging regular and complete voiding, avoiding prolonged periods of holding urine and emptying the bladder before and after sexual activity can reduce the risk of urinary tract infections.
Management of complicated urinary tract infections	Hospitalisation: Severe or complicated cases may require hospitalisation for intravenous antibiotics and close monitoring.
	Special populations: Management may differ for pregnant women, individuals with diabetes and those with anatomical abnormalities.

Source: Adapted from Hagler et al. (2023), Swales (2022), Yaqoob and Ashman (2021) and Jones and Stegall (2019).

Table 8.4 provides an overview of key aspects in the management of people with urinary tract infections.

RECURRENT URINARY TRACT INFECTION

The management of recurrent urinary tract infections involves a comprehensive and individualised approach. The aim is to prevent future episodes and to address underlying factors that may contribute to recurrence.

Identifying predisposing factors is a key step in the management process, encompassing anatomical abnormalities, urinary tract obstructions and conditions affecting immune function. Regular urinalysis and urine cultures are used to monitor for signs of infection and guide appropriate antibiotic therapy when needed.

For preventive measures, low-dose antibiotics over an extended period such as prophylaxis may be provided, especially for individuals with frequent and well-documented urinary tract infections. Patient education is a key preventative factor.

In certain cases, lifestyle modifications and immunomodulatory therapy (this involves interventions that aim to modify or regulate the immune response) may be considered. The identification and management of contributing factors, including underlying conditions such as diabetes or urinary tract obstructions, are integral to the management plan.

Regular follow-up appointments are required to monitor the patient's condition, assess the effectiveness of preventive measures and make necessary adjustments to the management plan. Referral to experts, such as urologists or infectious disease specialists, may be considered for complex cases or when additional expertise is needed.

The management of recurrent urinary tract infection involves a collaborative and ongoing partnership between healthcare providers and patients. The goal is to develop a tailored and effective management plan that addresses the unique challenges and factors associated with recurrent urinary tract infections.

HEALTH TEACHING

Patient education plays a crucial role in the management of urinary tract infections, emphasising the importance of medication adherence and recognising symptoms of recurrent infections. Healthcare providers must stay informed about the latest research and clinical guidelines to ensure that management approaches align with current best practices, promoting optimal outcomes for patients with urinary tract infections. Key health teaching needs for people with urinary tract infections are outlined in Table 8.5.

Table 8.5 Key health teaching needs for people with urinary tract infection

Intervention	Discussion
Understanding urinary tract infection causes	Provide information and advice about the common causes of urinary tract infections, including bacterial entry into the urinary tract and factors that may increase susceptibility.
Symptom recognition	Offer individuals information so they are able to recognise the signs and symptoms of urinary tract infections, such as pain or burning during urination, increased frequency, urgency, cloudy or malodorous urine and lower abdominal (suprapubic) discomfort.
Hygiene practices	Emphasise effective perineal hygiene, including wiping from front to back after bowel movements, to prevent the introduction of bacteria into the urethra.
Fluid intake	Encourage an adequate intake of fluids to promote frequent urination and help flush bacteria from the urinary tract.
Bladder emptying habits	Advise regular and complete bladder emptying, avoiding prolonged periods of holding urine and urinating before and after sexual activity to reduce the risk of bacterial colonisation.
Avoiding irritants	Offer information and advice to individuals on avoiding potential irritants, such as harsh soaps, bubble baths and personal hygiene sprays, which may contribute to urethral irritation.
Antibiotic adherence	Stress the importance of completing the full course of prescribed antibiotics, even if symptoms improve before the medication is finished, to ensure complete eradication of the infection.
Follow-up care	Discuss the importance of attending follow-up appointments to monitor the resolution of symptoms and assess the effectiveness of treatment (if required).

Intervention	Discussion
Preventive measures	Provide information on preventive measures, including lifestyle modifications such as maintaining a healthy diet, staying physically active and avoiding excessive use of certain contraceptives that may contribute to urinary tract infections.
When to seek medical attention	Offer information and advice to individuals on when to seek prompt medical attention, especially if symptoms worsen, new symptoms develop or if there are signs of systemic involvement such as fever, chills or back pain.
Psychosocial impact	Address the psychosocial impact of urinary tract infections, including any concerns or anxieties related to sexual activity and intimate relationships. Key points to consider: • Impact on emotional well-being • Body image and self-esteem • Sexual concerns • Communication challenges (individuals may find it challenging to communicate with their partners about the impact of a urinary tract infection on their intimate lives) • Relationship strain • Educational support • Psychological support (offering psychological support, such as counselling or access to mental health resources) • Holistic approach to care • Encouraging open dialogue • Incorporating partner support (if appropriate)

Source: Adapted from Swales (2022).

It is essential that health teaching needs are tailored to the specific needs and concerns of each individual with a urinary tract infection. By providing comprehensive education, healthcare providers can empower patients to actively participate in their care, reduce the likelihood of recurrences and maintain optimal urinary health.

CONCLUSION

Understanding urinary tract infection is important in understanding the details involved in the prevention, assessment and management of this common condition. This chapter has explored the aetiology, pathophysiology, clinical manifestations and various aspects of care related to people with urinary tract infections.

The importance of thorough patient history taking, recognising signs and symptoms and undertaking physical assessments are fundamental requirements in the early identification of urinary tract infection. The significance of timely and accurate diagnosis, using urinalysis and culture, cannot be overstated, as it guides the appropriate selection of antibiotics and influences the overall management plan.

An holistic approach to care is required, which considers not only the physical aspects but also the psychosocial impact on individuals. Understanding the emotional and relational challenges associated with urinary tract infection facilitates the provision of empathetic and patient-centred care, fostering open communication and collaboration with patients.

GLOSSARY OF TERMS

Antibiotic susceptibility testing: Laboratory test assessing the effectiveness of antibiotics against a specific bacterial strain.

Asymptomatic bacteriuria: Presence of bacteria in the urine without symptoms of infection; common in certain populations.

Bacteriuria: Presence of bacteria in the urine, a diagnostic criterion for urinary tract infections.

Catheter-associated urinary tract infection: Urinary tract infections that develop in individuals with indwelling urinary catheters.

Cystitis: Inflammation of the bladder, often resulting from a bacterial infection.

Dysuria: Painful or difficult urination, often a common symptom of urinary tract infections.

Fluoroquinolones: A class of antibiotics commonly used to treat urinary tract infections; includes ciprofloxacin and levofloxacin.

Frequency: Increased need to urinate, a common symptom of urinary tract infections.

Haematuria: Presence of blood in the urine, which can be a sign of infection or other urinary issues.

Midstream specimen of urine: Sometimes referred to as a 'clean-catch' urine specimen, it is a method of collecting urine for laboratory testing. Commonly used to obtain a sample less likely to be contaminated by bacteria or cells from the external genital area.

Pyelonephritis: Infection of the kidneys, which can be a serious and potentially severe form of urinary tract infection.

Pyuria: Presence of pus in the urine, indicating inflammation or infection.

Recurrent UTI: Multiple occurrences of urinary tract infections, often defined by a certain frequency within a specific time frame.

Urethritis: Inflammation of the urethra, commonly caused by infection.

Urinalysis: Laboratory tests examining physical and chemical properties of urine, often used to diagnose urinary tract infections.

Urinary tract infection: An infection affecting any part of the urinary system, including the kidneys, bladder, ureters and urethra.

Urine culture: Laboratory test to identify the specific bacteria causing the urinary tract infection and determine their susceptibility to antibiotics.

Urgency: Strong and sudden need to urinate, often associated with discomfort.

MULTIPLE CHOICE QUESTIONS

1. What is the most common causative agent of urinary tract infections (UTIs)?
   a) *Staphylococcus aureus*
   b) *Escherichia coli (E. coli)*

    c) *Streptococcus pyogenes*
    d) *Pseudomonas aeruginosa*

2. Which part of the urinary tract is commonly affected by cystitis?
   a) Kidneys
   b) Bladder
   c) Ureters
   d) Urethra

3. What is the primary symptom of dysuria in UTIs?
   a) Frequent urination
   b) Painful urination
   c) Blood in the urine
   d) Urgency to urinate

4. What is the preferred method for collecting a midstream specimen of urine for urinalysis?
   a) Collect the entire voided urine
   b) Collect the initial stream
   c) Collect the midstream portion
   d) Collect the final stream

5. What is the purpose of antibiotic susceptibility testing in UTIs?
   a) To identify the causative bacteria
   b) To determine the patient's immune response
   c) To assess the effectiveness of antibiotics
   d) To measure urinary pH

6. What is a common complication of untreated or recurrent UTIs that involves bacteria entering the bloodstream?
   a) Sepsis
   b) Meningitis
   c) Endocarditis
   d) Pneumonia

7. What is the primary purpose of encouraging increased fluid intake in UTI management?
   a) To reduce urinary urgency
   b) To flush bacteria from the urinary tract
   c) To alleviate dysuria
   d) To prevent haematuria

8. Which anatomical part of the female reproductive system contributes to the higher incidence of UTIs in women?
   a) Ovaries
   b) Uterus
   c) Vaginal cavity
   d) Fallopian tubes

9. Which of the following is a preventive measure to reduce the risk of UTIs?
   a) Holding urine for extended periods
   b) Wiping from front to back after bowel movements

c) Decreasing fluid intake

d) Avoiding regular voiding

10. Which population is more prone to catheter-associated urinary tract infections?
a) Pregnant women
b) Older adults
c) Children
d) Young adults

REFERENCES

Clare, C. (2021). Specimen collection (Chapter 14). In: *The Nursing Associate's Handbook of Clinical Skills* (ed. I. Peate). Oxford: Wiley.

European Association of Urology (2023). *EAU Guidelines on Urological Infections* (accessed December 2023).

Hagler, D., Harding, M.M., Kwong, J. et al. (2023). *Lewis's Medical-Surgical Nursing*, 12e. St Louis: Elsevier.

Jones, K. and Steggall, M. (2019). Nursing patients with urinary disorders (Chapter 9). In: *Alexander's Nursing Practice,* 5e. (ed. I. Peate). London: Elsevier.

Kaur, R. and Kaur, R. (2020). Symptoms, risk factors, diagnosis and treatment of urinary tract infection. *Postgraduate Medical Journal* 97: 803–812.

National Institute for Health and Care Excellence (2018). Urinary tract infection (recurrent): antimicrobial prescribing. https://www.nice.org.uk/guidance/ng112/resources/urinary-tract-infection-recurrent-antimicrobial-prescribing-pdf-66141595059397 (accessed December 2023).

National Institute for Health and Care Excellence (2023a). Urinary tract infection (lower) – men: how common is it? https://cks.nice.org.uk/topics/urinary-tract-infection-lower-men/background-information/prevalence/ (accessed December 2023).

National Institute for Health and Care Excellence (2023b). Urinary tract infection (lower) – women. https://cks.nice.org.uk/topics/urinary-tract-infection-lower-women/ (accessed December 2023).

National Institute for Health and Care Excellence (2023c). Urinary tract infection in adults. https://www.nice.org.uk/guidance/qs90/resources/urinary-tract-infections-in-adults-pdf-2098962322117 (accessed December 2023).

O'Callaghan, C. (2016). *The Renal System at a Glance,* 4e. Oxford: Wiley.

Public Health England (2020). Diagnosis of urinary tract infections. Quick reference tool for primary care for consultation and local adaptation. https://assets.publishing.service.gov.uk/government/uploads/system/uploads/attachment_data/file/927195/UTI_diagnostic_flowchart_NICE-October_2020-FINAL.pdf (accessed December 2023).

Swales, J. (2022). The person with a urinary disorder (Chapter 29). In: *Nursing Practice* 3e. (eds. I. Peate and A. Mitchell). Oxford: Wiley.

Yaqoob, M.M. and Ashman, M. (2021). Kidney and urinary tract disease (Chapter 36). In: *Kumar and Clark's Clinical Medicine,* 10e (eds. A. Feather, D. Randall and M. Waterhouse). London: Elsevier.

MCQ Answers

Chapter 1   Anatomy and Physiology: The Renal System

1. (b); 2. (a); 3. (c); 4. (b); 5. (c); 6. (b); 7. (a); 8. (b); 9. (c); 10. (b).

Chapter 2   Renal System Assessment

1. (b); 2. (b); 3. (a); 4. (a); 5. (c); 6. (c); 7. (d); 8. (b); 9. (a); 10. (d).

Chapter 3   Pyelonephritis

1. (b); 2. (b); 3. (b); 4. (c); 5. (c); 6. (b); 7. (d); 8. (b); 9. (b); 10. (d).

Chapter 4   Acute Kidney Injury

1. (b); 2. (c); 3. (b); 4. (b); 5. (b); 6. (a); 7. (a); 8. (c); 9. (b); 10. (d).

Chapter 5   Chronic Kidney Disease

1. (b); 2. (b); 3. (a); 4. (d); 5. (b); 6. (b); 7. (b); 8. (c); 9. (c); 10. (b).

Chapter 6   Renal Calculi

1. (c); 2. (c); 3. (b); 4. (b); 5. (b); 6. (c); 7. (b); 8. (c); 9. (b); 10. (c).

Chapter 7   Bladder Cancer

1. (b); 2. (b); 3. (c); 4. (b); 5. (c); 6. (c); 7. (c); 8. (a); 9. (d); 10. (b).

Chapter 8   Urinary Tract Infection

1. (b); 2. (b); 3. (b); 4. (c); 5. (c); 6. (a); 7. (b); 8. (c); 9. (b); 10. (b).

Note: Page numbers in *italics* and **bold** refer to figures and tables, respectively.